An Introduction to Acoustics

An Introduction to Acoustics

Edited by
Triston Rodriguez

☐ Larsen & Keller
www.larsen-keller.com

An Introduction to Acoustics
Edited by Triston Rodriguez
ISBN: 978-1-63549-013-8 (Hardback)

▤ Larsen & Keller

Published by Larsen and Keller Education,
5 Penn Plaza,
19th Floor,
New York, NY 10001, USA

Cataloging-in-Publication Data

An introduction to acoustics / edited by Triston Rodriguez.
    p. cm.
Includes bibliographical references and index.
ISBN 978-1-63549-013-8
1. Sound. 2. Sound--Transmission. 3. Acoustical engineering. I. Rodriguez, Triston.
QC225.15 .I58 2017
534--dc23

The publisher's policy is to use permanent paper from mills that operate a sustainable forestry policy. Furthermore, the publisher ensures that the text paper and cover boards used have met acceptable environmental accreditation standards.

Printed and bound in the United States of America.

For more information regarding Larsen and Keller Education and its products, please visit the publisher's website www.larsen-keller.com

Table of Contents

Preface

This book elucidates the concepts and innovative models around prospective developments with respect to acoustics. It describes in detail the various theories and practices related to this field. Acoustics refers to the scientific study of mechanical waves present in liquids, gases and solids. It deals with phenomena like infrasound, vibration, sound and ultrasound. The topics covered in this text are appropriate for students of acoustics. The book attempts to understand the multiple branches that fall under the discipline of acoustics and how such concepts have practical applications. This textbook is an essential guide for both students.

A short introduction to every chapter is written below to provide an overview of the content of the book:

Chapter 1 - Acoustics studies mechanical waves produced in matter. It includes subjects like sounds, vibrations and infrasound waves. This chapter will facilitate an integrated understanding of acoustics by providing comprehensive insight on the subject; **Chapter 2 -** Due to its multiple applications, acoustics is divided into various branches. The branches of acoustics explained to the reader in this chapter are aeroacoustics, architectural acoustics, musical acoustics, underwater acoustics and physical acoustics. Each of these contributing subdivisions is helpful in explaining the various facets of acoustics; **Chapter 3 -** To gain a meaningful understanding of acoustics, it is imperative to understand its basic concepts. This chapter discusses some of the important concepts like acoustic wave equation, fluid, P-wave and diffraction; **Chapter 4 -** The major aspects of acoustics are discussed in this chapter. Vibration, sound, ultrasound and infrasound are some of the significant topics related to acoustics. The following chapter unfolds its crucial aspects in a critical yet systematic manner; **Chapter 5 -** A device that converts one form of energy to another is a transducer. This chapter explains to the reader the importance of transducers in acoustics. There are several transduction devices in use on day-to-day basis, like loudspeakers, microphones and hydrophones. These devices have been explained in the following chapter; **Chapter 6 -** The force that exists between electrically charged particles is termed as electromagnetic force. Electromagnetism studies the phenomena of electromagnetic force. Transduction involves the application of certain principles like electromagnetism, electrostatics, piezoelectricity and sound reinforcement system. This section seeks to explain the fundamentals of transduction; **Chapter 7** The manner in which waves travel is studied as wave propagation. Some of the topics listed in this chapter are sound pressure, absolute threshold of hearing and interference. The chapter on wave propagation offers an insightful focus, keeping in mind the complex subject matter.

I extend my sincere thanks to the publisher for considering me worthy of this task. Finally, I thank my family for being a source of support and help.

Editor

Introduction to Acoustics

Acoustics studies mechanical waves produced in matter. It includes subjects like sounds, vibrations and infrasound waves. This chapter will facilitate an integrated understanding of acoustics by providing comprehensive insight on the subject.

Acoustics

Acoustics is the interdisciplinary science that deals with the study of all mechanical waves in gases, liquids, and solids including topics such as vibration, sound, ultrasound and infrasound. A scientist who works in the field of acoustics is an acoustician while someone working in the field of acoustics technology may be called an acoustical engineer. The application of acoustics is present in almost all aspects of modern society with the most obvious being the audio and noise control industries.

Artificial omni-directional sound source in an anechoic chamber

Hearing is one of the most crucial means of survival in the animal world, and speech is one of the most distinctive characteristics of human development and culture. Accordingly, the science of acoustics spreads across many facets of human society—music, medicine, architecture, industrial production, warfare and more. Likewise, animal species such as songbirds and frogs use sound and hearing as a key element of mating rituals or marking territories. Art, craft, science and technology have provoked one another to advance the whole, as in many other fields of knowledge. Robert Bruce Lindsay's 'Wheel of Acoustics' is a well accepted overview of the various fields in acoustics.

The Latin synonym is "sonic", after which the term sonics used to be a synonym for acoustics and later a branch of acoustics. Frequencies above and below the audible range are called "ultrasonic" and "infrasonic", respectively.

History

Early Research in Acoustics

In the 6th century BC, the ancient Greek philosopher Pythagoras wanted to know why some combinations of musical sounds seemed more beautiful than others, and he found answers in terms of numerical ratios representing the harmonic overtone series on a string. He is reputed to have observed that when the lengths of vibrating strings are expressible as ratios of integers (e.g. 2 to 3, 3 to 4), the tones produced will be harmonious, and the smaller the integers the more harmonious the sounds. If, for example, a string of a certain length would sound particularly harmonious with a string of twice the length (other factors being equal). In modern parlance, if a string sounds the note C when plucked, a string twice as long will sound a C an octave lower. In one system of musical tuning, the tones in between are then given by 16:9 for D, 8:5 for E, 3:2 for F, 4:3 for G, 6:5 for A, and 16:15 for B, in ascending order.

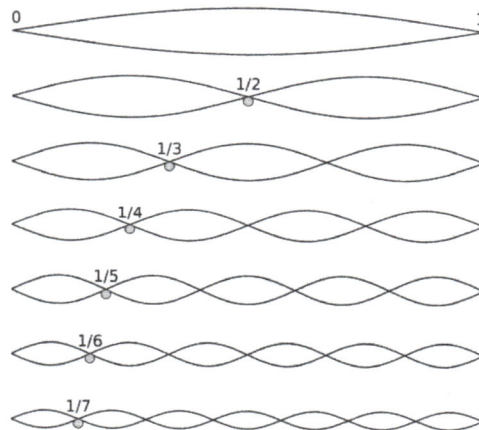

The fundamental and the first 6 overtones of a vibrating string. The earliest records of the study of this phenomenon are attributed to the philosopher Pythagoras in the 6th century BC.

Aristotle (384-322 BC) understood that sound consisted of compressions and rarefactions of air which "falls upon and strikes the air which is next to it...", a very good expression of the nature of wave motion.

In about 20 BC, the Roman architect and engineer Vitruvius wrote a treatise on the acoustic properties of theaters including discussion of interference, echoes, and reverberation—the beginnings of architectural acoustics. In Book V of his *De architectura* (*The Ten Books of Architecture*) Vitruvius describes sound as a wave comparable to a water wave extended to three dimensions, which, when interrupted by obstructions, would flow back and break up following waves. He described the ascending seats in ancient theaters as designed to prevent this deterioration of sound and also recommended bronze vessels of appropriate sizes be placed in theaters to resonate with the fourth, fifth and so on, up to the double octave, in order to resonate with the more desirable, harmonious notes.

Principles of acoustics have been applied since ancient times : A Roman theatre in the city of Amman.

The physical understanding of acoustical processes advanced rapidly during and after the Scientific Revolution. Mainly Galileo Galilei (1564–1642) but also Marin Mersenne (1588–1648), independently, discovered the complete laws of vibrating strings (completing what Pythagoras and Pythagoreans had started 2000 years earlier). Galileo wrote "Waves are produced by the vibrations of a sonorous body, which spread through the air, bringing to the tympanum of the ear a stimulus which the mind interprets as sound", a remarkable statement that points to the beginnings of physiological and psychological acoustics. Experimental measurements of the speed of sound in air were carried out successfully between 1630 and 1680 by a number of investigators, prominently Mersenne. Meanwhile, Newton (1642–1727) derived the relationship for wave velocity in solids, a cornerstone of physical acoustics (Principia, 1687).

Age of Enlightenment and Onward

The eighteenth century saw major advances in acoustics as mathematicians applied the new techniques of calculus to elaborate theories of sound wave propagation. In the nineteenth century the major figures of mathematical acoustics were Helmholtz in Germany, who consolidated the field of physiological acoustics, and Lord Rayleigh in England, who combined the previous knowledge with his own copious contributions to the field in his monumental work *The Theory of Sound* (1877). Also in the 19th century, Wheatstone, Ohm, and Henry developed the analogy between electricity and acoustics.

The twentieth century saw a burgeoning of technological applications of the large body of scientific knowledge that was by then in place. The first such application was Sabine's groundbreaking work in architectural acoustics, and many others followed. Underwater acoustics was used for detecting submarines in the first World War. Sound recording and the telephone played important roles in a global transformation of society. Sound measurement and analysis reached new levels of accuracy and sophistication through the use of electronics and computing. The ultrasonic frequency range enabled wholly new kinds of application in medicine and industry. New kinds of transducers (generators and receivers of acoustic energy) were invented and put to use.

Fundamental Concepts of Acoustics

Jay Pritzker Pavilion

At Jay Pritzker Pavilion, a LARES system is combined with a zoned sound reinforcement system, both suspended on an overhead steel trellis, to synthesize an indoor acoustic environment outdoors.

Definition

Acoustics is defined by ANSI/ASA S1.1-2013 as "(a) Science of sound, including its production, transmission, and effects, including biological and psychological effects. (b) Those qualities of a room that, together, determine its character with respect to auditory effects."

The study of acoustics revolves around the generation, propagation and reception of mechanical waves and vibrations.

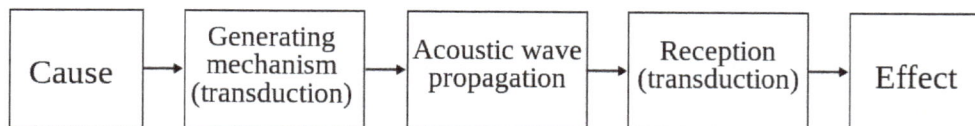

The steps shown in the above diagram can be found in any acoustical event or process. There are many kinds of cause, both natural and volitional. There are many kinds of transduction process that convert energy from some other form into sonic energy, producing a sound wave. There is one fundamental equation that describes sound wave propagation, the acoustic wave equation, but the phenomena that emerge from it are varied and often complex. The wave carries energy throughout the propagating medium. Eventually this energy is transduced again into other forms, in ways that again may be natural and/or volitionally contrived. The final effect may be purely physical or it may reach far into the biological or volitional domains. The five basic steps are found equally well whether we are talking about an earthquake, a submarine using sonar to locate its foe, or a band playing in a rock concert.

The central stage in the acoustical process is wave propagation. This falls within the domain of physical acoustics. In fluids, sound propagates primarily as a pressure wave. In solids, mechanical waves can take many forms including longitudinal waves, transverse waves and surface waves.

Acoustics looks first at the pressure levels and frequencies in the sound wave and how the wave interacts with the environment. This interaction can be described as either a diffraction, interference or a reflection or a mix of the three. If several media are present, a refraction can also occur. Transduction processes are also of special importance to acoustics.

Wave Propagation: Pressure Levels

In fluids such as air and water, sound waves propagate as disturbances in the ambient pressure level. While this disturbance is usually small, it is still noticeable to the human ear. The smallest sound that a person can hear, known as the threshold of hearing, is nine orders of magnitude smaller than the ambient pressure. The loudness of these disturbances is called the sound pressure level (SPL), and is measured on a logarithmic scale in decibels.

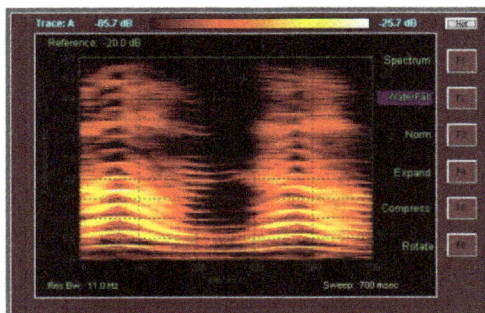

Spectrogram of a young girl saying "oh, no"

Wave Propagation: Frequency

Physicists and acoustic engineers tend to discuss sound pressure levels in terms of frequencies, partly because this is how our ears interpret sound. What we experience as "higher pitched" or "lower pitched" sounds are pressure vibrations having a higher or lower number of cycles per second. In a common technique of acoustic measurement, acoustic signals are sampled in time, and then presented in more meaningful forms such as octave bands or time frequency plots. Both of these popular methods are used to analyze sound and better understand the acoustic phenomenon.

The entire spectrum can be divided into three sections: audio, ultrasonic, and infrasonic. The audio range falls between 20 Hz and 20,000 Hz. This range is important because its frequencies can be detected by the human ear. This range has a number of applications, including speech communication and music. The ultrasonic range refers to the very high frequencies: 20,000 Hz and higher. This range has shorter wavelengths which allow better resolution in imaging technologies.

Medical applications such as ultrasonography and elastography rely on the ultrasonic frequency range. On the other end of the spectrum, the lowest frequencies are known as the infrasonic range. These frequencies can be used to study geological phenomena such as earthquakes.

Analytic instruments such as the spectrum analyzer facilitate visualization and measurement of acoustic signals and their properties. The spectrogram produced by such an instrument is a graphical display of the time varying pressure level and frequency profiles which give a specific acoustic signal its defining character.

Transduction in Acoustics

A transducer is a device for converting one form of energy into another. In an electroacoustic context, this means converting sound energy into electrical energy (or vice versa). Electroacoustic transducers include loudspeakers, microphones, hydrophones and sonar projectors. These devices convert a sound pressure wave to or from an electric signal. The most widely used transduction principles are electromagnetism, electrostatics and piezoelectricity.

An inexpensive low fidelity 3.5 inch driver, typically found in small radios

The transducers in most common loudspeakers (e.g. woofers and tweeters), are electromagnetic devices that generate waves using a suspended diaphragm driven by an electromagnetic voice coil, sending off pressure waves. Electret microphones and condenser microphones employ electrostatics—as the sound wave strikes the microphone's diaphragm, it moves and induces a voltage change. The ultrasonic systems used in medical ultrasonography employ piezoelectric transducers. These are made from special ceramics in which mechanical vibrations and electrical fields are interlinked through a property of the material itself.

Acoustician

An acoustician is an expert in the science of sound.

Education

There are many types of acoustician, but they usually have a Bachelor's degree or higher qualification. Some possess a degree in acoustics, while others enter the discipline via studies in fields such as physics or engineering. Much work in acoustics requires a good grounding in Mathematics and science. Many acoustic scientists work in research and development. Some conduct basic research to advance our knowledge of the perception (e.g. hearing, psychoacoustics or neurophysiology) of speech, music and noise. Other acoustic scientists advance understanding of how sound is affected as it moves through environments, e.g. Underwater acoustics, Architectural acoustics or Structural acoustics. Others areas of work are listed under subdisciplines below. Acoustic scientists work in government, university and private industry laboratories. Many go on to work in Acoustical Engineering. Some positions, such as Faculty (academic staff) require a Doctor of Philosophy.

Subdisciplines

These subdisciplines are a slightly modified list from the PACS (Physics and Astronomy Classification Scheme) coding used by the Acoustical Society of America.

Archaeoacoustics

The Divje Babe flute

Archaeoacoustics is the study of sound within archaeology. This typically involves studying the acoustics of archaeological sites and artefacts.

Aeroacoustics

Aeroacoustics is the study of noise generated by air movement, for instance via turbulence, and the movement of sound through the fluid air. This knowledge is applied in acoustical engineering to study how to quieten aircraft. Aeroacoustics is important to understanding how wind musical instruments work.

Acoustic Signal Processing

Acoustic signal processing is the electronic manipulation of acoustic signals. Applications include: active noise control; design for hearing aids or cochlear implants; echo cancellation; music information retrieval, and perceptual coding (e.g. MP3 or Opus).

Architectural Acoustics

Symphony Hall Boston where auditorium acoustics began

Architectural acoustics (also known as building acoustics) involves the scientific understanding of how to achieve a good sound within a building. It typically involves the study of speech intelligibility, speech privacy and music quality in the built environment.

Bioacoustics

Bioacoustics is the scientific study of the hearing and calls of animal calls, as well as how animals are affected by the acoustic and sounds of their habitat.

Electroacoustics

This subdiscipline is concerned with the recording, manipulation and reproduction of audio using electronics. This might include products such as mobile phones, large scale public address systems or virtual reality systems in research laboratories.

Environmental Noise and Soundscapes

Environmental acoustics is concerned with noise and vibration caused by railways, road traffic, aircraft, industrial equipment and recreational activities. The main aim of these studies is to reduce levels of environmental noise and vibration. Research work now also has a focus on the positive use of sound in urban environments: soundscapes and tranquility.

Musical Acoustics

Musical acoustics is the study of the physics of acoustic instruments; the audio signal processing used in electronic music; the computer analysis of music and composition, and the perception and cognitive neuroscience of music.

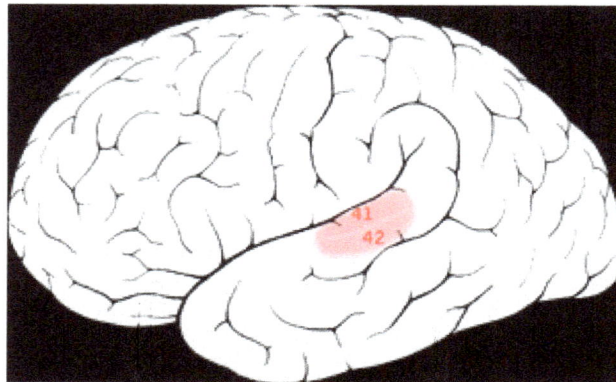

The primary auditory cortex is one of the main areas associated with superior pitch resolution.

Speech

Acousticians study the production, processing and perception of speech. Speech recognition and Speech synthesis are two important areas of speech processing using computers. The subject also overlaps with the disciplines of physics, physiology, psychology, and linguistics.

Ultrasonics

Ultrasonics deals with sounds at frequencies too high to be heard by humans. Specialisms include medical ultrasonics (including medical ultrasonography), sonochemistry, material characterisation and underwater acoustics (Sonar).

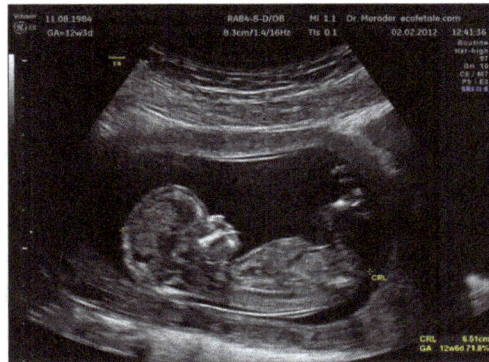

Ultrasound image of a fetus in the womb, viewed at 12 weeks of pregnancy (bidimensional-scan)

Underwater Acoustics

Underwater acoustics is the scientific study of natural and man-made sounds underwater. Applications include sonar to locate submarines, underwater communication by whales, climate change monitoring by measuring sea temperatures acoustically, sonic weapons, and marine bioacoustics.

Vibration and Dynamics

This is the study of how mechanical systems vibrate and interact with their surroundings. Applications might include: ground vibrations from railways; vibration isolation to reduce vibration in operating theatres; studying how vibration can damage health (vibration white finger); vibration control to protect a building from earthquakes, or measuring how structure-borne sound moves through buildings.

Mechanical Wave

A mechanical wave is a wave that is an oscillation of matter, and therefore transfers energy through a medium. While waves can move over long distances, the movement of the medium of transmission—the material—is limited. Therefore, oscillating material does not move far from its initial equilibrium position. Mechanical waves transport energy. This energy propagates in the same direction as the wave. Any kind of wave (mechanical or electromagnetic) has a certain energy. Mechanical waves can be produced only in media which possess elasticity and inertia.

Ripple in water is a surface wave.

A mechanical wave requires an initial energy input. Once this initial energy is added, the wave travels through the medium until all its energy is transferred. In contrast, electromagnetic waves require no medium, but can still travel through one.

One important property of mechanical waves is that their amplitudes are measured in an unusual way, displacement divided by (reduced) wavelength. When this gets comparable to unity, significant nonlinear effects such as harmonic generation may occur, and, if large enough, may result in chaotic effects. For example, waves on the surface of a body of water break when this dimensionless amplitude exceeds 1, resulting in a foam on the surface and turbulent mixing. Some of the most common examples of mechanical waves are water waves, sound waves, and seismic waves.

There are three types of mechanical waves: transverse waves, longitudinal waves, and surface waves.

Transverse Wave

Transverse waves cause the medium to vibrate at a right angle to the direction of the wave or energy being carried by the medium. Transverse waves have two parts—the crest and the trough. The crest is the highest point of the wave and the trough is the lowest. The distance between a crest and a trough is half of wavelength. The wavelength is the distance from crest to crest or from trough to trough.

To see an example, move an end of a Slinky (whose other end is fixed) to the left-and-right of the Slinky (as opposed to-and-fro the Slinky). Light also has properties of a transverse wave, although it is an electromagnetic wave.

Longitudinal Wave

Longitudinal waves cause the medium to vibrate parallel to the direction of the wave. It consists of multiple compressions and rarefactions. The rarefaction is the farthest distance apart in the longitudinal wave and the compression is the closest distance together. The speed of the longitudinal wave is increased in higher index of refraction, due to the closer proximity of the atoms in the medium that is being compressed. Sound is considered a longitudinal wave.

Surface Waves

This type of wave travels along a surface that is between two media. An example of a surface wave would be waves in a pool, or in an ocean, lake, or any other type of water body. There are two types of surface waves, namely Rayleigh waves and Love waves.

Rayleigh waves, also known as *ground roll*, are waves that travel as ripples with motion similar to those of waves on the surface of water. Rayleigh waves are much slower than body waves, roughly 90% of the velocity of body waves for a typical homogeneous elastic medium.

A Love wave is a surface waves having horizontal waves that are shear or transverse to the direction of propagation. They usually travel slightly faster than Rayleigh waves, about 90% of the body wave velocity, and have the largest amplitude.

Forms of Mechanical Waves

Longitudinal Wave

Longitudinal waves, also known as "1 waves", are waves in which the displacement of the medium is in the same direction as, or the opposite direction to, the direction of travel of the wave. Mechanical longitudinal waves are also called compressional waves or compression waves, because they produce compression and rarefaction when traveling through a medium, and pressure waves, because they produce increases and decreases in pressure. The other main type of wave is the transverse wave, in which the displacements of the medium are at right angles to the direction of propagation. Some transverse waves are mechanical, meaning that the wave needs a medium to travel through. Transverse mechanical waves are also called "t-waves" or "shear waves".

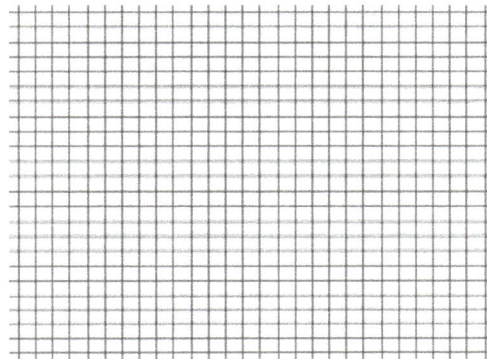

Plane pressure pulse wave

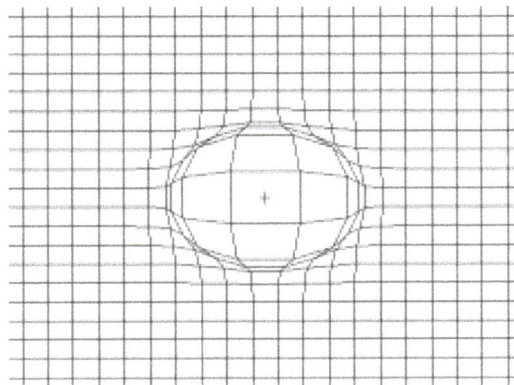

Representation of the propagation of an omnidirectional pulse wave on a 2d grid (empirical shape)

Examples

Longitudinal waves include sound waves (vibrations in pressure, particle of displacement, and particle velocity propagated in an elastic medium) and seismic P-waves (created by earthquakes and explosions). In longitudinal waves, the displacement of the medium is parallel to the propagation of the wave. A wave along the length of a stretched Slinky toy, where the distance between coils increases and decreases, is a good visualization. Sound waves in air are longitudinal, pressure waves.

Sound Waves

In the case of longitudinal harmonic sound waves, the frequency and wavelength can be described by the formula

$$y(x,t) = y_0 \cos\left(\omega\left(t - \frac{x}{c}\right)\right)$$

where:

- y is the displacement of the point on the traveling sound wave;
- x is the distance the point has traveled from the wave's source;
- t is the time elapsed;
- y_0 is the amplitude of the oscillations,
- c is the speed of the wave; and
- ω is the angular frequency of the wave.

The quantity x/c is the time that the wave takes to travel the distance x.

The ordinary frequency (f) of the wave is given by

$$f = \frac{\omega}{2\pi}.$$

For sound waves, the amplitude of the wave is the difference between the pressure of the undisturbed air and the maximum pressure caused by the wave.

Sound's propagation speed depends on the type, temperature, and composition of the medium through which it propagates.

Pressure Waves

In an elastic medium with rigidity, a harmonic pressure wave oscillation has the form,

$$y(x,t) = y_0 \cos(kx - \omega t + \varphi)$$

where:

- y_0 is the amplitude of displacement,
- k is the wavenumber,
- x is the distance along the axis of propagation,
- ω is the angular frequency,
- t is the time, and
- φ is the phase difference.

The restoring force, which acts to return the medium to its original position, is provided by the medium's bulk modulus.

Electromagnetic

Maxwell's equations lead to the prediction of electromagnetic waves in a vacuum, which are transverse (in that the electric fields and magnetic fields vary perpendicularly to the direction of propagation). However, waves can exist in plasmas or confined spaces, called plasma waves, which can be longitudinal, transverse, or a mixture of both. Plasma waves can also occur in force-free magnetic fields.

In the early development of electromagnetism, there were some like Alexandru Proca (1897-1955) known for developing relativistic quantum field equations bearing his name (Proca's equations) for the massive, vector spin-1 mesons. In recent decades some extended electromagnetic theorists, such as Jean-Pierre Vigier and Bo Lehnert of the Swedish Royal Society, have used the Proca equation in an attempt to demonstrate photon mass as a longitudinal electromagnetic component of Maxwell's equations, suggesting that longitudinal electromagnetic waves could exist in a Dirac polarized vacuum.

After Heaviside's attempts to generalize Maxwell's equations, Heaviside came to the conclusion that electromagnetic waves were not to be found as longitudinal waves in *"free space"* or homogeneous media. But Maxwell's equations do lead to the appearance of longitudinal waves under some circumstances, for example, in plasma waves or guided waves. Basically distinct from the "free-space" waves, such as those studied by Hertz in his UHF experiments, are Zenneck waves. The longitudinal modes of a resonant cavity are the particular standing wave patterns formed by waves confined in a cavity. The longitudinal modes correspond to those wavelengths of the wave which are reinforced by constructive interference after many reflections from the cavity's reflecting surfaces. Recently, Haifeng Wang et al. proposed a method that can generate a longitudinal electromagnetic (light) wave in free space, and this wave can propagate without divergence for a few wavelengths.

Transverse Wave

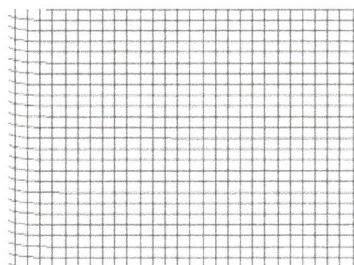

Transverse plane wave

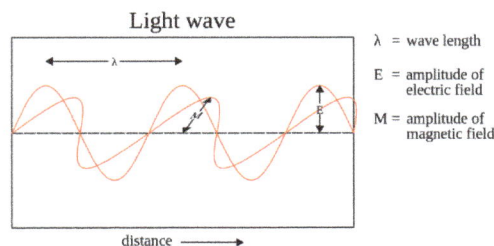

A light wave is an example of a transverse wave.

A transverse wave is a moving wave that consists of oscillations occurring perpendicular (or right angled) to the direction of energy transfer. If a transverse wave is moving in the positive x-direction, its oscillations are in up and down directions that lie in the y–z plane. Light is an example of a transverse wave. With regard to transverse waves in matter, the displacement of the medium is perpendicular to the direction of propagation of the wave. A ripple in a pond and a wave on a string are easily visualized as transverse waves.

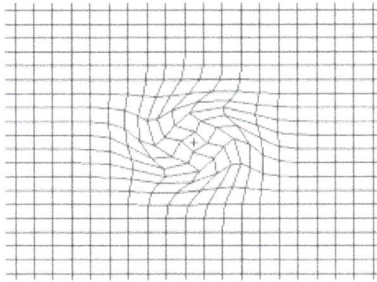

Propagation of a transverse spherical wave in a 2d grid (empirical model)

Explanation

Transverse waves are waves that are oscillating perpendicularly to the direction of propagation. If you anchor one end of a ribbon or string and hold the other end in your hand, you can create transverse waves by moving your hand up and down. Notice though, that you can also launch waves by moving your hand side-to-side. This is an important point. There are two independent directions in which wave motion can occur. In this case, these motions are the y and z directions mentioned above, while the wave propagates away in the x direction.

'Polarized' Waves

Continuing with the string example, if you carefully move your hand in a clockwise circle, you will launch waves in the form of a left-handed helix as they propagate away. Similarly, if you move your hand in a counter-clockwise circle, a right-handed helix will form. These phenomena of *simultaneous* motion in two directions go beyond the kinds of waves you can create on the surface of water; in general a wave on a string can be *two-dimensional*. Two-dimensional transverse waves exhibit a phenomenon called polarization. A wave produced by moving your hand in a line, up and down for instance, is a linearly polarized wave, a special case. A wave produced by moving your hand in a circle is a circularly polarized wave, another special case. If your motion is not strictly in a line or a circle your hand will describe an ellipse and the wave will be elliptically polarized.

Electromagnetic Waves

Electromagnetic waves behave in this same way, although it is slightly harder to see. Electromagnetic waves are also two-dimensional transverse waves. Ray theory does not describe phenomena such as interference and diffraction, which require wave theory (involving the phase of the wave). You can think of a ray of light, in optics, as an idealized narrow beam of electromagnetic radiation. Rays are used to model the propagation of light through an optical system, by dividing the real light field up into discrete rays that can be computationally propagated through the system by the techniques of ray tracing. A light ray is a line or curve that is perpendicular to the light's wave-fronts (and is therefore collinear with the wave vector). Light rays bend at the interface between two dissimilar media and may be curved in a medium in which the refractive index changes. Geometric optics describes how rays propagate through an optical system.

This two-dimensional nature should not be confused with the two components of an electromagnetic wave, the electric and magnetic field components, which are shown in the electromagnetic wave diagram here. The light wave diagram shows linear polarization. Each of these fields, the electric and the magnetic, exhibits two-dimensional transverse wave behavior, just like the waves on a string.

The transverse plane wave animation shown is also an example of linear polarization. The wave shown could occur on a water surface.

Surface Wave

In physics, a surface wave is a mechanical wave that propagates along the interface between differing media, usually as a gravity wave between two fluids with different densities. A surface wave can also be an elastic (or a seismic) wave, such as with a *Rayleigh* or *Love* wave. It can also be an electromagnetic wave guided by a refractive index gradient. In radio transmission, a ground wave is a surface wave that propagates close to the surface of the Earth.

Diving grebe creates surface waves

Mechanical Waves

In seismology, several types of surface waves are encountered. Surface waves, in this mechanical sense, are commonly known as either *Love waves* (L waves) or *Rayleigh waves*. A seismic wave is a wave that *travels through the Earth, often as the result of an earthquake or explosion*. Love waves have transverse motion (movement is perpendicular to the direction of travel, like light waves), whereas Rayleigh waves have both longitudinal (movement parallel to the direction of travel, like sound waves) and transverse motion. Seismic waves are studied by seismologists and measured by a seismograph or seismometer. Surface waves span a wide frequency range, and the period of waves that are most damaging is usually 10 seconds or longer. Surface waves can travel around the globe many times from the largest earthquakes. Surface waves are caused when P waves and S waves come to the surface.

The term "surface wave" can describe waves over an ocean, even when they are approximated by Airy functions and are more properly called creeping waves. Examples are the waves at the surface of water and air (ocean surface waves), or ripples in the sand at the interface with water or air. Another example is internal waves, which can be transmitted along the interface of two water masses of different densities.

In theory of Hearing physiology, the Traveling Wave (TW) of Von Bekesy, resulted from an acoustic surface wave of the basilar membrane into the cochlear duct. His theory pretended to explain every features of the auditory sensation owing to these passive mechanical phenomena. But Jozef Zwislocki and later David Kemp, showed that that was irrealistic and that an active feedback was necessary.

Electromagnetic Waves

Ground waves refer to the propagation of radio waves parallel to and adjacent to the surface of the Earth, following the curvature of the Earth. These *surface waves* are also known loosely as the Norton surface wave, the Zenneck surface wave, Sommerfeld waves, and gliding waves.

Radio Propagation

Lower frequencies, below 3 MHz, travel efficiently as ground waves. In ITU nomenclature, this includes (in order): medium frequency (MF), low frequency (LF), very low frequency (VLF), ultra low frequency (ULF), super low frequency (SLF), extremely low frequency (ELF) waves.

Ground propagation works because lower-frequency waves are more strongly diffracted around obstacles due to their long wavelengths, allowing them to follow the Earth's curvature. The Earth has one refractive index and the atmosphere has another, thus constituting an interface that supports the surface wave transmission. Ground waves propagate in vertical polarization, with their magnetic field horizontal and electric field (close to) vertical. With VLF waves, the Ionosphere and earth's surface act as a waveguide.

Conductivity of the surface affects the propagation of ground waves, with more conductive surfaces such as sea water providing better propagation. Increasing the conductivity in a surface results in less dissipation. The refractive indices are subject to spatial and temporal changes. Since the ground is not a perfect electrical conductor, ground waves are attenuated as they follow the earth's surface. The wavefronts initially are vertical, but the ground, acting as a lossy dielectric, causes the wave to tilt forward as it travels. This directs some of the energy into the earth where it is dissipated, so that the signal decreases exponentially.

Most long-distance LF "longwave" radio communication (between 30 kHz and 300 kHz) is a result of groundwave propagation. Mediumwave radio transmissions (frequencies between 300 kHz and 3000 kHz), including AM broadcast band, travel both as groundwaves and, for longer distances at night, as skywaves. Ground losses become lower at lower frequencies, greatly increasing the coverage of AM stations using the lower end of the band. The VLF and LF frequencies are mostly used for military communications, especially with ships and submarines. The lower the frequency the better the waves penetrate sea water. ELF waves (below 3 kHz) have even been used to communicate with deeply submerged submarines.

Surface waves have been used in over-the-horizon radar, which operates mainly at frequencies between 2 and 20 MHz over the sea, which has a sufficiently high conductivity to convey the surface waves to and from a reasonable distance (up to 100 km or more; over-horizon radar also uses skywave propagation at much greater distances). In the development of radio, surface waves were used extensively. Early commercial and professional radio services relied exclusively on long wave, low frequencies and ground-wave propagation. To prevent interference with these services, amateur and experimental transmitters were restricted to the high frequencies (HF), felt to be useless since their ground-wave range was limited. Upon discovery of the other propagation modes possible at medium wave and short wave frequencies, the advantages of HF for commercial and military purposes became apparent. Amateur experimentation was then confined only to authorized frequencies in the range.

Mediumwave and shortwave reflect off the ionosphere at night, which is known as skywave. During daylight hours, the lower D layer of the ionosphere forms and absorbs lower frequency energy. This prevents skywave propagation from being very effective on mediumwave frequencies in daylight hours. At night, when the D layer dissipates, mediumwave transmissions travel better by skywave. Ground waves *do not* include ionospheric and tropospheric waves.

The propagation of sound waves through the ground taking advantage of the earths ability to more efficiently transmit low frequency is known as Audio ground wave (AGW).

Microwave Field Theory

Within microwave field theory, the interface of a dielectric and conductor supports "surface wave transmission". Surface waves have been studied as part of transmission lines and some may be considered as single-wire transmission lines.

Characteristics and utilizations of the electrical surface wave phenomenon include:

- The field components of the wave diminish with distance from the interface.

- Electromagnetic energy is not converted from the surface wave field to another form of energy (except in leaky or lossy surface waves) such that the wave does not transmit power normal to the interface, i.e. it is evanescent along that dimension.

- In optical fiber transmission, evanescent waves are surface waves.

- In coaxial cable in addition to the TEM mode there also exists a transverse-magnetic (TM) mode which propagates as a surface wave in the region around the central conductor. For coax of common impedance this mode is effectively suppressed but in high impedance coax and on a single central conductor without any outer shield, low attenuation and very broadband propagation is supported. Transmission line operation in this mode is called E-Line.

References

- Scarre, Christopher (2006). Archaeoacoustics. McDonald Institute for Archaeological Research. ISBN 978-1902937359.

- da Silva, Andrey Ricardo (2009). Aeroacoustics of Wind Instruments: Investigations and Numerical Methods. VDM Verlag. ISBN 978-3639210644.

- Templeton, Duncan (1993). Acoustics in the Built Environment: Advice for the Design Team. Architectural Press. ISBN 978-0750605380.

- Krylov, V.V. (Ed.) (2001). Noise and Vibration from High-speed Trains. Thomas Telford. ISBN 9780727729637.

- World Health Organisation (2011). Burden of disease from environmental noise (PDF). WHO. ISBN 978 92 890 0229 5.

- Pohlmann, Ken (2010). Principles of Digital Audio, Sixth Edition. McGraw Hill Professional. p. 336. ISBN 9780071663472.

- D. Lohse, B. Schmitz & M. Versluis (2001). "Snapping shrimp make flashing bubbles". Nature. 413 (6855): 477–478. Bibcode:2001Natur.413..477L. doi:10.1038/35097152. PMID 11586346.

- Structural Acoustics & Vibration Technical Committee. "Structural Acoustics & Vibration Technical Committee". Retrieved 22 May 2013.

- Ernst Mach, Introduction to The Science of Mechanics: A Critical and Historical Account of its Development (1893, 1960) Tr. Thomas J. McCormack

- Lakes, R. (1998). Experimental limits on the photon mass and cosmic magnetic vector potential. Physical review letters, 80(9), 1826-1829

- Heaviside, Oliver, "Electromagnetic theory". Appendices: D. On compressional electric or magnetic waves. Chelsea Pub Co; 3rd edition (1971) 082840237X

- Corum, K. L., and J. F. Corum, "The Zenneck surface wave", Nikola Tesla, Lightning observations, and stationary waves, Appendix II. 1994.

- Haifeng Wang, Luping Shi, Boris Luk'yanchuk, Colin Sheppard and Chong Tow Chong, "Creation of a needle of longitudinally polarized light in vacuum using binary optics," Nature Photonics, Vol.2, pp 501-505, 2008, doi:10.1038/nphoton.2008.127.

- This article incorporates public domain material from the General Services Administration document "Federal Standard 1037C" (in support of MIL-STD-188).

- Dyakonov, M. I. (April 1988). "New type of electromagnetic wave propagating at an interface". Soviet Physics JETP. 67 (4): 714.

Branches of Acoustics

Due to its multiple applications, acoustics is divided into various branches. The branches of acoustics explained to the reader in this chapter are aeroacoustics, architectural acoustics, musical acoustics, underwater acoustics and physical acoustics. Each of these contributing subdivisions is helpful in explaining the various facets of acoustics.

Aeroacoustics

Aeroacoustics is a branch of acoustics that studies noise generation via either turbulent fluid motion or aerodynamic forces interacting with surfaces. Noise generation can also be associated with periodically varying flows. A notable example of this phenomenon are the Aeolian tones produced by wind blowing over fixed objects.

Although no complete scientific theory of the generation of noise by aerodynamic flows has been established, most practical aeroacoustic analysis relies upon the so-called *aeroacoustic analogy*, proposed by Sir James Lighthill in the 1950s while at the University of Manchester. whereby the governing equations of motion of the fluid are coerced into a form reminiscent of the wave equation of "classical" (i.e. linear) acoustics in the left-hand side with the remaining terms as sources in the right-hand side.

History

The modern discipline of aeroacoustics can be said to have originated with the first publication of Lighthill in the early 1950s, when noise generation associated with the jet engine was beginning to be placed under scientific scrutiny.

Lighthill's Equation

Lighthill rearranged the Navier–Stokes equations, which govern the flow of a compressible viscous fluid, into an inhomogeneous wave equation, thereby making a connection between fluid mechanics and acoustics. This is often called "Lighthill's analogy" because it presents a model for the acoustic field that is not, strictly speaking, based on the physics of flow-induced/generated noise, but rather on the analogy of how they might be represented through the governing equations of a compressible fluid.

The first equation of interest is the conservation of mass equation, which reads

$$\frac{\partial \rho}{\partial t} + \nabla \cdot \left(\rho \mathbf{v} \right) = \frac{D\rho}{Dt} + \rho \nabla \cdot \mathbf{v} = 0,$$

where ρ and $\mathbf{v}$ represent the density and velocity of the fluid, which depend on space and time, and D/Dt is the substantial derivative.

Next is the conservation of momentum equation, which is given by

$$\rho\frac{\partial \mathbf{v}}{\partial t}+\rho(\mathbf{v}\cdot\nabla)\mathbf{v}=-\nabla p+\nabla\cdot\sigma,$$

where p is the thermodynamic pressure, and σ is the viscous (or traceless) part of the stress tensor from the Navier–Stokes equations.

Now, multiplying the conservation of mass equation by $\mathbf{v}$ and adding it to the conservation of momentum equation gives

$$\frac{\partial}{\partial t}(\rho\mathbf{v})+\nabla\cdot(\rho\mathbf{v}\otimes\mathbf{v})=-\nabla p+\nabla\cdot\sigma.$$

Note that $\mathbf{v}\otimes\mathbf{v}$ is a tensor. Differentiating the conservation of mass equation with respect to time, taking the divergence of the conservation of momentum equation and subtracting the latter from the former, we arrive at

$$\frac{\partial^2\rho}{\partial t^2}-\nabla^2 p+\nabla\cdot\nabla\cdot\sigma=\nabla\cdot\nabla\cdot(\rho\mathbf{v}\otimes\mathbf{v}).$$

Subtracting $c_0^2\nabla^2\rho$, where c_0 is the speed of sound in the medium in its equilibrium (or quiescent) state, from both sides of the last equation and rearranging it results in

$$\frac{\partial^2\rho}{\partial t^2}-c_0^2\nabla^2\rho=\nabla\cdot\left[\nabla\cdot(\rho\mathbf{v}\otimes\mathbf{v})-\nabla\cdot\sigma+\nabla p-c_0^2\nabla\rho\right],$$

which is equivalent to

$$\frac{\partial^2\rho}{\partial t^2}-c_0^2\nabla^2\rho=(\nabla\otimes\nabla):\left[\rho\mathbf{v}\otimes\mathbf{v}-\sigma+(p-c_0^2\rho)\mathbb{I}\right],$$

where $\mathbb{I}$ is the identity tensor, and $:$ denotes the (double) tensor contraction operator.

The above equation is the celebrated Lighthill equation of aeroacoustics. It is a wave equation with a source term on the right-hand side, i.e. an inhomogeneous wave equation. The argument of the "double-divergence operator" on the right-hand side of last equation, i.e. $\rho\mathbf{v}\otimes\mathbf{v}-\sigma+(p-c_0^2\rho)\mathbb{I}$, is the so-called *Lighthill turbulence stress tensor for the acoustic field*, and it is commonly denoted by T.

Using Einstein notation, Lighthill's equation can be written as

$$\frac{\partial^2\rho}{\partial t^2}-c_0^2\nabla^2\rho=\frac{\partial^2 T_{ij}}{\partial x_i\partial x_j},\quad(*)$$

where

$$T_{ij}=\rho v_i v_j-\sigma_{ij}+(p-c_0^2\rho)\delta_{ij},$$

and δ_{ij} is the Kronecker delta. Each of the acoustic source terms, i.e. terms in T_{ij}, may play a significant role in the generation of noise depending upon flow conditions considered. $\rho v_i v_j$ σ_{ij} describes unsteady convection of flow (or Reynolds' Stress, developed by Osborne Reynolds), σ_{ij} describes sound generated by shear, and $(p-c_0^2\rho)\delta_{ij}$ describes non-linear acoustic generation processes.

In practice, it is customary to neglect the effects of viscov sity on the fluid, i.e. one takes $\sigma=0$, because it is generally accepted that the effects of the latter on noise generation, in most situations, are orders of magnitude smaller than those due to the other terms. Lighthill provides an in-depth discussion of this matter.

In aeroacoustic studies, both theoretical and computational efforts are made to solve for the acoustic source terms in Lighthill's equation in order to make statements regarding the relevant aerodynamic noise generation mechanisms present.

Finally, it is important to realize that Lighthill's equation is exact in the sense that no approximations of any kind have been made in its derivation.

Related Model Equations

In their classical text on fluid mechanics, Landau and Lifshitz derive an aeroacoustic equation analogous to Lighthill's (i.e., an equation for sound generated by "turbulent" fluid motion), but for the incompressible flow of an inviscid fluid. The inhomogeneous wave equation that they obtain is for the *pressure* p rather than for the density ρ of the fluid. Furthermore, unlike Lighthill's equation, Landau and Lifshitz's equation is not exact; it is an approximation.

If one is to allow for approximations to be made, a simpler way (without necessarily assuming the fluid is incompressible) to obtain an approximation to Lighthill's equation is to assume that $p - p_0 = c_0^2(\rho - \rho_0)$, where ρ_0 and p_0 are the (characteristic) density and pressure of the fluid in its equilibrium state. Then, upon substitution the assumed relation between pressure and density into (*) we obtain the equation

$$\frac{1}{c_0^2}\frac{\partial^2 p}{\partial t^2} - \nabla^2 p = \frac{\partial^2 \tilde{T}_{ij}}{\partial x_i \partial x_j}, \quad \text{where} \quad \tilde{T}_{ij} = \rho v_i v_j.$$

And for the case when the fluid is indeed incompressible, i.e. $\rho = \rho_0$ (for some positive constant ρ_0) everywhere, then we obtain exactly the equation given in Landau and Lifshitz, namely

$$\frac{1}{c_0^2}\frac{\partial^2 p}{\partial t^2} - \nabla^2 p = \rho_0 \frac{\partial^2 \hat{T}_{ij}}{\partial x_i \partial x_j}, \quad \text{where} \quad \hat{T}_{ij} = v_i v_j.$$

A similar approximation [in the context of equation (*)], namely $T \approx \rho_0 \hat{T}$, is suggested by Lighthill.

Of course, one might wonder whether we are justified in assuming that $p - p_0 = c_0^2(\rho - \rho_0)..$ The answer is affirmative, if the flow satisfies certain basic assumptions. In particular, if $\rho \ll \rho_0$ and $p \ll p_0$, then the assumed relation follows directly from the *linear* theory of sound waves In fact, the approximate relation between p and ρ that we assumed is just a linear approximation to the generic barotropic equation of state of the fluid.

However, even after the above deliberations, it is still not clear whether one is justified in using an inherently *linear* relation to simplify a *nonlinear* wave equation. Nevertheless, it is a very common practice in nonlinear acoustics as the textbooks on the subject show: e.g., Naugolnykh and Ostrovsky and Hamilton and Morfey.

Architectural Acoustics

Architectural acoustics (also known as room acoustics and building acoustics) is the science and engineering of achieving a good sound within a building and is a branch of acoustical engineering. The first application of modern scientific methods to architectural acoustics was carried out by Wallace Sabine in the Fogg Museum lecture room who then applied his new found knowledge to the design of Symphony Hall, Boston.

Symphony Hall, Birmingham, an example of the application of architectural acoustics.

Architectural acoustics can be about achieving good speech intelligibility in a theatre, restaurant or railway station, enhancing the quality of music in a concert hall or recording studio, or suppressing noise to make offices and homes more productive and pleasant places to work and live in. Architectural acoustic design is usually done by acoustic consultants.

Building Skin Envelope

This science analyzes noise transmission from building exterior envelope to interior and vice versa. The main noise paths are roofs, eaves, walls, windows, door and penetrations. Sufficient control ensures space functionality and is often required based on building use and local municipal codes. An example would be providing a suitable design for a home which is to be constructed close to a high volume roadway, or under the flight path of a major airport, or of the airport itself.

Inter-Space Noise Control

The science of limiting and/or controlling noise transmission from one building space to another to ensure space functionality and speech privacy. The typical sound paths are ceilings, room partitions, acoustic ceiling panels (such as wood dropped ceiling panels), doors, windows, flanking, ducting and other penetrations. Technical solutions depend on the source of the noise and the path of acoustic transmission, for example noise by steps or noise by (air, water) flow vibrations. An example would be providing suitable party wall design in an apartment complex to minimize the mutual disturbance due to noise by residents in adjacent apartments.

Interior Space Acoustics

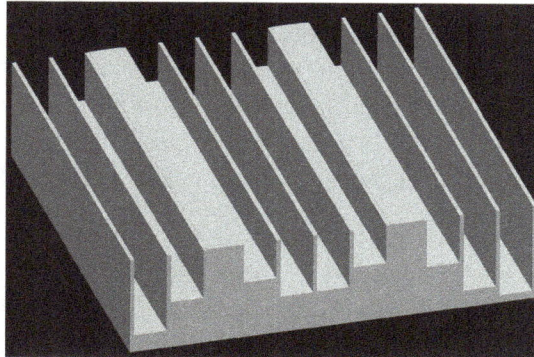

Diffusers which scatter sound are used in some rooms to improve the acoustics

This is the science of controlling a room's surfaces based on sound absorbing and reflecting properties. Excessive reverberation time, which can be calculated, can lead to poor speech intelligibility.

Ceiling of Culture Palace (Tel Aviv) concert hall is covered with perforated metal panels

Sound reflections create standing waves that produce natural resonances that can be heard as a pleasant sensation or an annoying one. Reflective surfaces can be angled and coordinated to provide good coverage of sound for a listener in a concert hall or music recital space. To illustrate this concept consider the difference between a modern large office meeting room or lecture theater and a traditional classroom with all hard surfaces.

An anechoic chamber, using acoustic absorption to create a "dead" space.

Interior building surfaces can be constructed of many different materials and finishes. Ideal acoustical panels are those without a face or finish material that interferes with the acoustical infill or substrate. Fabric covered panels are one way to heighten acoustical absorption. Perforated metal shows also sound absorbing qualities. Finish material is used to cover over the acoustical substrate. Mineral fiber board, or Micore, is a commonly used acoustical substrate. Finish materials often consist of fabric, wood or acoustical tile. Fabric can be wrapped around substrates to create what is referred to as a "pre-fabricated panel" and often provides good noise absorption if laid onto a wall.

Prefabricated panels are limited to the size of the substrate ranging from 2 by 4 feet (0.61 m × 1.22 m) to 4 by 10 feet (1.2 m × 3.0 m). Fabric retained in a wall-mounted perimeter track system, is referred to as "on-site acoustical wall panels". This is constructed by framing the perimeter track into shape, infilling the acoustical substrate and then stretching and tucking the fabric into the perimeter frame system. On-site wall panels can be constructed to accommodate door frames, baseboard, or any other intrusion. Large panels (generally, greater than 50 square feet (4.6 m²)) can be created on walls and ceilings with this method. Wood finishes can consist of punched or routed slots and provide a natural look to the interior space, although acoustical absorption may not be great.

There are three ways to improve workplace acoustics and solve workplace sound problems – the ABCs.

- A = Absorb (via drapes, carpets, ceiling tiles, etc.)

- B = Block (via panels, walls, floors, ceilings and layout)

- C = Cover-up (via sound masking)

Mechanical Equipment Noise

Building services noise control is the science of controlling noise produced by:

- ACMV (air conditioning and mechanical ventilation) systems in buildings, termed HVAC in North America

- Elevators

- Electrical generators positioned within or attached to a building

- Any other building service infrastructure component that emits sound.

Inadequate control may lead to elevated sound levels within the space which can be annoying and reduce speech intelligibility. Typical improvements are vibration isolation of mechanical equipment, and sound traps in ductwork. Sound masking can also be created by adjusting HVAC noise to a predetermined level.

Musical Acoustics

Musical acoustics or music acoustics is the branch of acoustics concerned with researching and describing the physics of music – how sounds are employed to make music. Examples of areas of study are the function of musical instruments, the human voice (the physics of speech and sing-ing), computer analysis of melody, and in the clinical use of music in music therapy.

Methods and Fields of Study

- The physics of musical instruments

- Frequency range of music

- Frequency analysis

- Computer analysis of musical structure

- Synthesis of musical sounds

- Music cognition, based on physics (also known as psychoacoustics)

Physical Aspects

Whenever two different pitches are played at the same time, their sound waves interact with each other – the highs and lows in the air pressure reinforce each other to produce a different sound wave. As a result, any given sound wave which is not a sine wave can be modeled by many different sine waves of the appropriate frequencies and amplitudes (a frequency spectrum). In humans the hearing apparatus (composed of the ears and brain) can usually isolate these tones and hear them distinctly. When two or more tones are played at once, a variation of air pressure at the ear "contains" the pitches of each, and the ear and/or brain isolate and decode them into distinct tones.

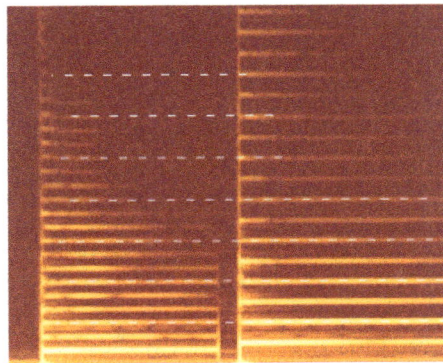

A spectrogram of a violin playing a note and then a perfect fifth above it. The shared partials are highlighted by the white dashes.

When the original sound sources are perfectly periodic, the note consists of several related sine waves (which mathematically add to each other) called the fundamental and the harmonics, par-tials, or overtones. The sounds have harmonic frequency spectra. The lowest frequency present is the fundamental, and is the frequency at which the entire wave vibrates. The overtones vibrate faster than the fundamental, but must vibrate at integer multiples of the fundamental frequency in order for the total wave to be exactly the same each cycle. Real instruments are close to peri-odic, but the frequencies of the overtones are slightly imperfect, so the shape of the wave changes slightly over time.

Subjective Aspects

Variations in air pressure against the ear drum, and the subsequent physical and neurological processing and interpretation, give rise to the subjective experience called *sound*. Most sound that people recognize as musical is dominated by periodic or regular vibrations rather than non-periodic ones; that is, musical sounds typically have a definite pitch). The transmission of these variations through air is via a sound wave. In a very simple case, the sound of a sine wave, which is considered to be the most basic model of a sound waveform, causes the air pressure to increase and decrease in a regular fashion, and is heard as a very pure tone. Pure tones can be produced by tuning forks or whistling. The rate at which the air pressure oscillates is the frequency of the tone, which is measured in oscillations per second, called hertz. Frequency is the primary determinant of the perceived pitch. Frequency of musical instruments can change with altitude due to changes in air pressure.

Harmonics, Partials, and Overtones

The fundamental is the frequency at which the entire wave vibrates. Overtones are other sinusoidal components present at frequencies above the fundamental. All of the frequency components that make up the total waveform, including the fundamental and the overtones, are called partials. Together they form the harmonic series.

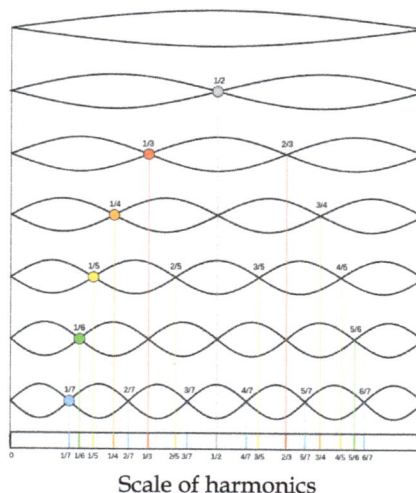

Scale of harmonics

Overtones which are perfect integer multiples of the fundamental are called harmonics. When an overtone is near to being harmonic, but not exact, it is sometimes called a harmonic partial, although they are often referred to simply as harmonics. Sometimes overtones are created that are not anywhere near a harmonic, and are just called partials or inharmonic overtones.

The fundamental frequency is considered the *first harmonic* and the *first partial*. The numbering of the partials and harmonics is then usually the same; the second partial is the second harmonic, etc. But if there are inharmonic partials, the numbering no longer coincides. Overtones are numbered as they appear *above* the fundamental. So strictly speaking, the *first* overtone is the *second* partial (and usually the *second* harmonic). As this can result in confusion, only harmonics are usually referred to by their numbers, and overtones and partials are described by their relationships to those harmonics.

Harmonics and Non-linearities

When a periodic wave is composed of a fundamental and only odd harmonics (f, 3f, 5f, 7f, ...), the summed wave is *half-wave symmetric*; it can be inverted and phase shifted and be exactly the same. If the wave has any even harmonics (0f, 2f, 4f, 6f, ...), it will be asymmetrical; the top half will not be a mirror image of the bottom.

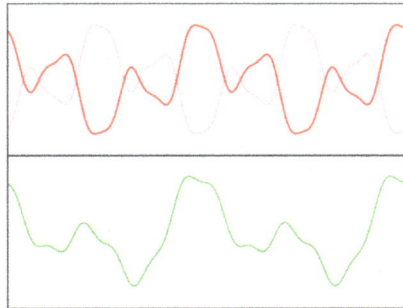

A symmetric and asymmetric waveform. The red (upper) wave contains only the fundamental and odd harmonics; the green (lower) wave contains the fundamental and even harmonics.

Conversely, a system which changes the shape of the wave (beyond simple scaling or shifting) creates additional harmonics (harmonic distortion). This is called a *non-linear system*. If it affects the wave symmetrically, the harmonics produced will only be odd, if asymmetrically, at least one even harmonic will be produced (and probably also odd).

Harmony

If two notes are simultaneously played, with frequency ratios that are simple fractions (e.g. 2/1, 3/2 or 5/4), then the composite wave will still be periodic with a short period, and the combination will sound consonant. For instance, a note vibrating at 200 Hz and a note vibrating at 300 Hz (a perfect fifth, or 3/2 ratio, above 200 Hz) will add together to make a wave that repeats at 100 Hz: every 1/100 of a second, the 300 Hz wave will repeat thrice and the 200 Hz wave will repeat twice. Note that the total wave repeats at 100 Hz, but there is not actually a 100 Hz sinusoidal component present.

Additionally, the two notes will have many of the same partials. For instance, a note with a fundamental frequency of 200 Hz will have harmonics at:

        (200,) 400, 600, 800, 1000, 1200, ...

A note with fundamental frequency of 300 Hz will have harmonics at:

        (300,) 600, 900, 1200, 1500, ...

The two notes share harmonics at 600 and 1200 Hz, and more will coincide further up the series.

The combination of composite waves with short fundamental frequencies and shared or closely related partials is what causes the sensation of harmony.

When two frequencies are near to a simple fraction, but not exact, the composite wave cycles slowly enough to hear the cancellation of the waves as a steady pulsing instead of a tone. This is called beating, and is considered to be unpleasant, or dissonant.

The frequency of beating is calculated as the difference between the frequencies of the two notes. For the example above, |200 Hz - 300 Hz| = 100 Hz. As another example, a combination of 3425 Hz and 3426 Hz would beat once per second (|3425 Hz - 3426 Hz| = 1 Hz). This follows from modulation theory.

The difference between consonance and dissonance is not clearly defined, but the higher the beat frequency, the more likely the interval to be dissonant. Helmholtz proposed that maximum dissonance would arise between two pure tones when the beat rate is roughly 35 Hz.

Scales

The material of a musical composition is usually taken from a collection of pitches known as a scale. Because most people cannot adequately determine absolute frequencies, the identity of a scale lies in the ratios of frequencies between its tones (known as intervals).

The diatonic scale appears in writing throughout history, consisting of seven tones in each octave. In just intonation the diatonic scale may be easily constructed using the three simplest intervals within the octave, the perfect fifth (3/2), perfect fourth (4/3), and the major third (5/4). As forms of the fifth and third are naturally present in the overtone series of harmonic resonators, this is a very simple process.

The following table shows the ratios between the frequencies of all the notes of the just major scale and the fixed frequency of the first note of the scale.

C	D	E	F	G	A	B	C
1	9/8	5/4	4/3	3/2	5/3	15/8	2

There are other scales available through just intonation, for example the minor scale. Scales which do not adhere to just intonation, and instead have their intervals adjusted to meet other needs are known as temperaments, of which equal temperament is the most used. Temperaments, though they obscure the acoustical purity of just intervals often have other desirable properties, such as a closed circle of fifths.

Nonlinear Acoustics

Nonlinear acoustics (NLA) is a branch of physics and acoustics dealing with sound waves of sufficiently large amplitudes. Large amplitudes require using full systems of governing equations of fluid dynamics (for sound waves in liquids and gases) and elasticity (for sound waves in solids). These equations are generally nonlinear, and their traditional linearization is no longer possible.

The solutions of these equations show that, due to the effects of nonlinearity, sound waves are being distorted as they travel.

Introduction

A sound wave propagates through a material as a localized pressure change. Increasing the pressure of a gas or fluid increases its local temperature. The local speed of sound in a compressible material increases with temperature; as a result, the wave travels faster during the high pressure phase of the oscillation than during the lower pressure phase. This affects the wave's frequency structure; for example, in an initially plane sinusoidal wave of a single frequency, the peaks of the wave travel faster than the troughs, and the pulse becomes cumulatively more like a saw-tooth wave. In other words, the wave self-distorts. In doing so, other frequency components are introduced, which can be described by the Fourier series. This phenomenon is characteristic of a non-linear system, since a linear acoustic system responds only to the driving frequency. This always occurs but the effects of geometric spreading and of absorption usually overcome the self distortion, so linear behavior usually prevails and nonlinear acoustic propagation occurs only for very large amplitudes and only near the source.

Additionally, waves of different amplitudes will generate different pressure gradients, contributing to the non-linear effect.

Physical Analysis

The pressure changes within a medium cause the wave energy to transfer to higher harmonics. Since attenuation generally increases with frequency, a counter effect exists that changes the nature of the nonlinear effect over distance. To describe their level of nonlinearity, materials can be given a nonlinearity parameter, B/A. The values of A and B are the coefficients of the first and second order terms of the Taylor series expansion of the equation relating the material's pressure to its density. The Taylor series has more terms, and hence more coefficients (C, D, .. etc.) but they are seldom used. Typical values for the nonlinearity parameter in biological mediums are shown in the following table.

Material	$B >$
Blood	6.1
Brain	6.6
Fat	10
Liver	6.8
Muscle	7.4
Water	5.2
Monoatomic Gas	0.67

In a liquid usually a modified coefficient is used known as $\beta = 1 + \dfrac{B}{2A}$.

Mathematical Model

Governing Equations to Derive Westervelt Equation

Continuity:

$$\frac{\partial \rho}{\partial t} + \nabla \cdot (\rho \mathbf{u}) = 0$$

Conservation of momentum:

$$\rho \left(\frac{\partial \mathbf{u}}{\partial t} + \mathbf{u} \cdot \nabla \mathbf{u} \right) + \nabla p = (\lambda + 2\mu) \nabla (\nabla \cdot \mathbf{u})$$

with Taylor perturbation expansion on density:

$$\rho = \sum_0^\infty \varepsilon^i \rho_i$$

where ε is a small parameter, i.e. the perturbation parameter, the equation of state becomes:

$$p = \varepsilon \rho_1 c_0^2 \left(1 + \varepsilon \frac{B}{2!A} \frac{\rho_1}{\rho_0} + O(\varepsilon^2) \right)$$

If the second term in the Taylor expansion of pressure is dropped, the viscous wave equation can be derived. If it is kept, the non-linear term in pressure appears in the Westervelt equation.

Westervelt Equation

The general wave equation that accounts for nonlinearity up to the second-order is given by the Westervelt equation

$$\nabla^2 p - \frac{1}{c_0^2} \frac{\partial^2 p}{\partial t^2} + \frac{\delta}{c_0^4} \frac{\partial^3 p}{\partial t^3} = -\frac{\beta}{\rho_0 c_0^4} \frac{\partial^2 p^2}{\partial t^2}$$

where p is the sound pressure, c_0 is the small signal sound speed, δ is the sound diffusivity, β is the non-linearity coefficient and ρ_0 is the ambient density.

The sound diffusivity is given by

$$\delta = \frac{1}{\rho_0} \left(\frac{4}{3}\mu + \mu_B \right) + \frac{k}{\rho_0} \left(\frac{1}{c_v} - \frac{1}{c_p} \right)$$

where μ is the shear viscosity, μ_B the bulk viscosity, k the thermal conductivity, c_v and c_p the specific heat at constant volume and pressure respectively.

Burgers' Equation

The Westervelt equation can be simplified to take a one-dimensional form with an assumption of strictly forward propagating waves and the use of a coordinate transformation to a retarded time frame:

$$c_p \frac{\partial p}{\partial z} - \frac{\beta}{\rho_0 c_0^3} p \frac{\partial p}{\partial \tau} = \frac{\delta}{2c_0^3} \frac{\partial^2 p}{\partial \tau^2}$$

where $\tau = t - z / c_0$ is retarded time. This corresponds to a viscous Burgers equation:

$$\frac{\partial y}{\partial t'} + u \frac{\partial y}{\partial x} = d \frac{\partial^2 y}{\partial x^2} ..$$

in the pressure field (y=p), with a mathematical "time variable":

$$t' = \frac{z}{c_0}..$$

and with a "space variable":

$$x = -\frac{\rho_0 c_0^2}{\beta}\tau..$$

and a negative diffusion coefficient:

$$d = -\frac{\rho_0 c_0}{2\beta^2}\delta..$$

The Burgers equation is the simplest equation that describes the combined effects of nonlinearity and losses on the propagation of progressive waves.

Kzk Equation

An augmentation to the Burgers equation that accounts for the combined effects of non-linearity, diffraction and absorption in directional sound beams is described by the Khokhlov-Zabolotskaya-Kuznetsov (KZK) equation. Solutions to this equation are generally used to model non-linear acoustics.

If the z axis is in the direction of the sound beam path and the (x, y) plane is perpendicular to that, the KZK equation can be written

$$\frac{\partial^2 p}{\partial z \partial \tau} = \frac{c_0}{2}\nabla_\perp^2 p + \frac{\delta}{2c_0^3}\frac{\partial^3 p}{\partial \tau^3} + \frac{\beta}{2\rho_0 c_0^3}\frac{\partial^2 p^2}{\partial \tau^2}$$

The equation can be solved for a particular system using a finite difference scheme. Such solutions show how the sound beam distorts as it passes through a non-linear medium.

Common Occurrences

Sonic Boom

The nonlinear behavior of the atmosphere leads to change of the wave shape in a sonic boom. Generally, this makes the boom more 'sharp' or sudden, as the high-amplitude peak moves to the wavefront.

Acoustic Levitation

The practice of acoustic levitation would not be possible without understanding nonlinear acoustic phenomena. The nonlinear effects are particularly evident due to the high-powered acoustic waves involved.

Ultrasonic Waves

Because of their relatively high amplitude to wavelength ratio, ultrasonic waves commonly display nonlinear propagation behavior. For example, nonlinear acoustics is a field of interest for medical ultrasonography because it can be exploited to produce better image quality.

Musical Acoustics

The physical behavior of musical acoustics in mainly nonlinear. Many attempts are made to model their sound generation from physical modeling of emulating their sound from measurements of their non-linearity.

Bioacoustics

Bioacoustics is a cross-disciplinary science that combines biology and acoustics. Usually it refers to the investigation of sound production, dispersion and reception in animals (including humans). This involves neurophysiological and anatomical basis of sound production and detection, and relation of acoustic signals to the medium they disperse through. The findings provide clues about the evolution of acoustic mechanisms, and from that, the evolution of animals that employ them.

In underwater acoustics and fisheries acoustics the term is also used to mean the effect of plants and animals on sound propagated underwater, usually in reference to the use of sonar technology for biomass estimation.

History

For a long time humans have employed animal sounds to recognise and find them. Bioacoustics as a scientific discipline was established by the Slovene biologist Ivan Regen who began systematically to study insect sounds. In 1925 he used a special stridulatory device to play in a duet with an insect. Later he put a male cricket behind a microphone and female crickets in front of a loudspeaker. The females were not moving towards the male but towards the loudspeaker. Regen's most important contribution to the field apart from realization that insects also detect airborne sounds was the discovery of tympanal organ's function.

Relatively crude electro-mechanical devices available at the time (such as phonographs) allowed only for crude appraisal of signal properties. More accurate measurements were made possible in the second half of the 20th century by advances in electronics and utilization of devices such as oscilloscopes and digital recorders.

The most recent advances in bioacoustics concern the relationships among the animals and their acoustic environment and the impact of anthropogenic noise. Bioacoustic techniques have recently been proposed as a non-invasive method for estimating biodiversity.

Methods in Bioacoustics

Listening is still one of the main methods used in bioacoustical research. Little is known about neurophysiological processes that play a role in production, detection and interpretation of sounds in animals, so animal behaviour and the signals themselves are used for gaining insight into these processes.

Hydrophone

Acoustic Signals

An experienced observer can use animal sounds to recognize a "singing" animal species, its location and condition in nature. Investigation of animal sounds also includes signal recording with electronic recording equipment. Due to the wide range of signal properties and media they propagate through, specialized equipment may be required instead of the usual microphone, such as a hydrophone (for underwater sounds), detectors of ultrasound (very high-frequency sounds) or infrasound (very low-frequency sounds), or a laser vibrometer (substrate-borne vibrational signals). Computers are used for storing and analysis of recorded sounds. Specialized sound-editing software is used for describing and sorting signals according to their intensity, frequency, duration and other parameters.

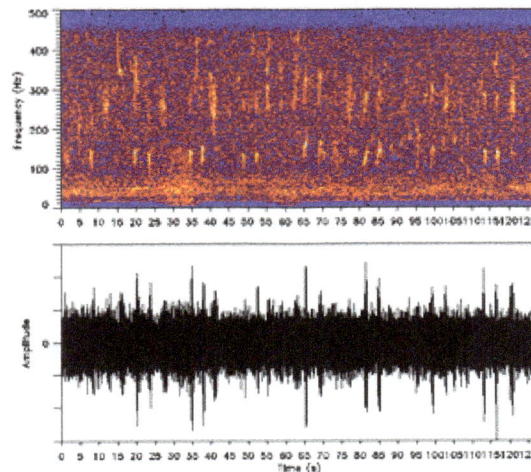

Spectrogram (above) and oscillogram (below) of the humpback whale's calls

Animal sound collections, managed by museums of natural history and other institutions, are an important tool for systematic investigation of signals. Many effective automated methods involving signal processing, data mining and machine learning techniques have been developed to detect and classify the bioacoustic signals.

Sound Production, Detection, and Use in Animals

Scientists in the field of bioacoustics are interested in anatomy and neurophysiology of organs involved in sound production and detection, including their shape, muscle action, and activity of neuronal networks involved. Of special interest is coding of signals with action potentials in the latter.

But since the methods used for neurophysiological research are still fairly complex and understanding of relevant processes is incomplete, more trivial methods are also used. Especially useful is observation of behavioural responses to acoustic signals. One such response is phonotaxis – directional movement towards the signal source. By observing response to well defined signals in a controlled environment, we can gain insight into signal function, sensitivity of the hearing apparatus, noise filtering capability, etc.

Biomass Estimation

Biomass estimation is a method of detecting and quantifying fish and other marine organisms using sonar technology. As the sound pulse travels through water it encounters objects that are of different density than the surrounding medium, such as fish, that reflect sound back toward the sound source. These echoes provide information on fish size, location, and abundance. The basic components of the scientific echo sounder hardware function is to transmit the sound, receive, filter and amplify, record, and analyze the echoes. While there are many manufacturers of commercially available "fish-finders," quantitative analysis requires that measurements be made with calibrated echo sounder equipment, having high signal-to-noise ratios.

Animal Sounds

European starling singing

Sounds used by animals that fall within the scope of bioacoustics include a wide range of frequencies and media, and are often not "*sound*" in the narrow sense of the word (i.e. compression waves that propagate through air and are detectable by the human ear). Katydid crickets, for example, communicate by sounds with frequencies higher than 100 kHz, far into the ultrasound range. Lower, but still in ultrasound, are sounds used by bats for echolocation. On the other side of the frequency spectrum are low frequency-vibrations, often not detected by hearing organs, but with other, less specialized sense organs. The examples include ground vibrations produced by elephants whose principal frequency component is around 15 Hz, and low- to medium-frequency substrate-borne vibrations used by most insect orders. Many animal sounds, however, do fall within the frequency range detectable by a human ear, between 20 and 20,000 Hz. Mechanisms for sound production and detection are just as diverse as the signals themselves.

Underwater Acoustics

Underwater acoustics is the study of the propagation of sound in water and the interaction of the mechanical waves that constitute sound with the water and its boundaries. The water may be in the ocean, a lake or a tank. Typical frequencies associated with underwater acoustics are between 10 Hz and 1 MHz. The propagation of sound in the ocean at frequencies lower than 10 Hz is usually not possible without penetrating deep into the seabed, whereas frequencies above 1 MHz are rarely used because they are absorbed very quickly. Underwater acoustics is sometimes known as hydroacoustics.

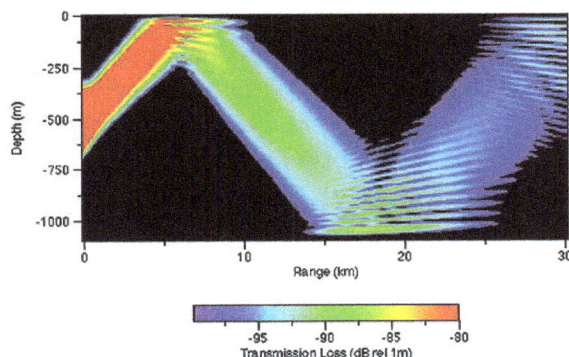

Output of a computer model of underwater acoustic propagation in a simplified ocean environment.

The field of underwater acoustics is closely related to a number of other fields of acoustic study, including sonar, transduction, acoustic signal processing, acoustical oceanography, bioacoustics, and physical acoustics.

History

Underwater sound has probably been used by marine animals for millions of years. The science of underwater acoustics began in 1490, when Leonardo da Vinci wrote the following,

> "If you cause your ship to stop and place the head of a long tube in the water and place the outer extremity to your ear, you will hear ships at a great distance from you."

In 1687 Isaac Newton wrote his Mathematical Principles of Natural Philosophy which included the first mathematical treatment of sound. The next major step in the development of underwater acoustics was made by Daniel Colladon, a Swiss physicist, and Charles Sturm, a French mathematician. In 1826, on Lake Geneva, they measured the elapsed time between a flash of light and the sound of a submerged ship's bell heard using an underwater listening horn. They measured a sound speed of 1435 metres per second over a 17 kilometre(Km) distance, providing the first quantitative measurement of sound speed in water. The result they obtained was within about 2% of currently accepted values. In 1877 Lord Rayleigh wrote the Theory of Sound and established modern acoustic theory.

The sinking of *Titanic* in 1912 and the start of World War I provided the impetus for the next wave of progress in underwater acoustics. Systems for detecting icebergs and U-boats were developed. Between 1912 and 1914, a number of echolocation patents were granted in Europe and the U.S., culminating in Reginald A. Fessenden's echo-ranger in 1914. Pioneering work was carried out during this time in France by Paul Langevin and in Britain by A B Wood and associates. The development of both active ASDIC and passive sonar (SOund Navigation And Ranging) proceeded apace during the war, driven by the first large scale deployments of submarines. Other advances in underwater acoustics included the development of acoustic mines.

In 1919, the first scientific paper on underwater acoustics was published, theoretically describing the refraction of sound waves produced by temperature and salinity gradients in the ocean. The range predictions of the paper were experimentally validated by transmission loss measurements.

The next two decades saw the development of several applications of underwater acoustics. The fathometer, or depth sounder, was developed commercially during the 1920s. Originally natural materials were used for the transducers, but by the 1930s sonar systems incorporating piezoelectric transducers made from synthetic materials were being used for passive listening systems and for active echo-ranging systems. These systems were used to good effect during World War II by both submarines and anti-submarine vessels. Many advances in underwater acoustics were made which were summarised later in the series Physics of Sound in the Sea, published in 1946.

After World War II, the development of sonar systems was driven largely by the Cold War, resulting in advances in the theoretical and practical understanding of underwater acoustics, aided by computer-based techniques.

Theory

Sound Waves in Water

A sound wave propagating underwater consists of alternating compressions and rarefactions of the water. These compressions and rarefactions are detected by a receiver, such as the human ear or a hydrophone, as changes in pressure. These waves may be man-made or naturally generated.

Speed of Sound, Density and Impedance

The speed of sound c (i.e., the longitudinal motion of wavefronts) is related to frequency f and wavelength λ of a wave by $c = f \cdot \lambda$.

This is different from the particle velocity u, which refers to the motion of molecules in the medium due to the sound, and relates the plane wave pressure p to the fluid density ρ and sound speed c by $p = c \cdot u \cdot \rho$.

The product of c and ρ from the above formula is known as the characteristic acoustic impedance. The acoustic power (energy per second) crossing unit area is known as the intensity of the wave and for a plane wave the average intensity is given by $I = q^2 / (\rho c)$, where q is the root mean square acoustic pressure.

At 1 kHz, the wavelength in water is about 1.5 m. Sometimes the term "sound velocity" is used but this is incorrect as the quantity is a scalar.

The large impedance contrast between air and water (the ratio is about 3600) and the scale of surface roughness means that the sea surface behaves as an almost perfect reflector of sound at frequencies below 1 kHz. Sound speed in water exceeds that in air by a factor of 4.4 and the density ratio is about 820.

Absorption of Sound

Absorption of low frequency sound is weak. The main cause of sound attenuation in fresh water, and at high frequency in sea water (above 100 kHz) is viscosity. Important additional contributions at lower frequency in seawater are associated with the ionic relaxation of boric acid (up to c. 10 kHz) and magnesium sulfate (c. 10 kHz-100 kHz).

Sound may be absorbed by losses at the fluid boundaries. Near the surface of the sea losses can occur in a bubble layer or in ice, while at the bottom sound can penetrate into the sediment and be absorbed.

Sound Reflection and Scattering

Boundary Interactions

Both the water surface and bottom are reflecting and scattering boundaries.

Surface

For many purposes the sea-air surface can be thought of as a perfect reflector. The impedance contrast is so great that little energy is able to cross this boundary. Acoustic pressure waves reflected from the sea surface experience a reversal in phase, often stated as either a "pi phase change" or a "180 deg phase change". This is represented mathematically by assigning a reflection coefficient of minus 1 instead of plus one to the sea surface.

At high frequency (above about 1 kHz) or when the sea is rough, some of the incident sound is scattered, and this is taken into account by assigning a reflection coefficient whose magnitude is less than one. For example, close to normal incidence, the reflection coefficient becomes $R = -e^{-2k^2 h^2 sin^2 A}$, where h is the rms wave height.

A further complication is the presence of wind generated bubbles or fish close to the sea surface. The bubbles can also form plumes that absorb some of the incident and scattered sound, and scatter some of the sound themselves.

Seabed

The acoustic impedance mismatch between water and the bottom is generally much less than at the surface and is more complex. It depends on the bottom material types and depth of the layers. Theories have been developed for predicting the sound propagation in the bottom in this case, for example by Biot and by Buckingham.

At Target

The reflection of sound at a target whose dimensions are large compared with the acoustic wavelength depends on its size and shape as well as the impedance of the target relative to that of water. Formulae have been developed for the target strength of various simple shapes as a function of angle of sound incidence. More complex shapes may be approximated by combining these simple ones.

Propagation of Sound

Underwater acoustic propagation depends on many factors. The direction of sound propagation is determined by the sound speed gradients in the water. In the sea the vertical gradients are generally much larger than the horizontal ones. Combining this with a tendency towards increasing sound speed at increasing depth, due to the increasing pressure in the deep sea, causes a reversal of the sound speed gradient in the thermocline, creating an efficient waveguide at the depth, corresponding to the minimum sound speed. The sound speed profile may cause regions of low sound intensity called "Shadow Zones," and regions of high intensity called "Caustics". These may be found by ray tracing methods.

At equator and temperate latitudes in the ocean, the surface temperature is high enough to reverse the pressure effect, such that a sound speed minimum occurs at depth of a few hundred metres. The presence of this minimum creates a special channel known as Deep Sound Channel, previously known as the SOFAR (sound fixing and ranging) channel, permitting guided propagation of underwater sound for thousands of kilometres without interaction with the sea surface or the seabed. Another phenomenon in the deep sea is the formation of sound focusing areas, known as Convergence Zones. In this case sound is refracted downward from a near-surface source and then back up again. The horizontal distance from the source at which this occurs depends on the positive and negative sound speed gradients. A surface duct can also occur in both deep and moderately shallow water when there is upward refraction, for example due to cold surface temperatures. Propagation is by repeated sound bounces off the surface.

In general, as sound propagates underwater there is a reduction in the sound intensity over increasing ranges, though in some circumstances a gain can be obtained due to focusing. *Propagation loss* (sometimes referred to as *transmission loss*) is a quantitative measure of the reduction in sound intensity between two points, normally the sound source and a distant receiver. If I_s is the far field intensity of the source referred to a point 1 m from its acoustic centre and I_r is the intensity at the receiver, then the propagation loss is given by $PL = 10log(I_s / I_r)$. In this equation I_r is not the true acoustic intensity at the receiver, which is a vector quantity, but a scalar equal to the equivalent plane wave intensity (EPWI) of the sound field. The EPWI is defined as the magnitude of the intensity of a plane wave of the same RMS pressure as the true acoustic field. At short range the propagation loss is dominated by spreading while at long range it is dominated by absorption and/or scattering losses.

An alternative definition is possible in terms of pressure instead of intensity, giving $PL = 20log(p_s / p_r)$, where p_s is the RMS acoustic pressure in the far-field of the projector, scaled to a standard distance of 1 m, and p_r is the RMS pressure at the receiver position.

These two definitions are not exactly equivalent because the characteristic impedance at the receiver may be different from that at the source. Because of this, the use of the intensity definition leads to a different sonar equation to the definition based on a pressure ratio. If the source and receiver are both in water, the difference is small.

Propagation Modelling

The propagation of sound through water is described by the wave equation, with appropriate boundary conditions. A number of models have been developed to simplify propagation calculations. These models include ray theory, normal mode solutions, and parabolic equation simplifications of the wave equation. Each set of solutions is generally valid and computationally efficient in a limited frequency and range regime, and may involve other limits as well. Ray theory is more appropriate at short range and high frequency, while the other solutions function better at long range and low frequency. Various empirical and analytical formulae have also been derived from measurements that are useful approximations.

Reverberation

Transient sounds result in a decaying background that can be of much larger duration than the original transient signal. The cause of this background, known as reverberation, is partly due to scattering from rough boundaries and partly due to scattering from fish and other biota. For an acoustic signal to be detected easily, it must exceed the reverberation level as well as the background noise level.

Doppler Shift

If an underwater object is moving relative to an underwater receiver, the frequency of the received sound is different from that of the sound radiated (or reflected) by the object. This change in frequency is known as a Doppler shift. The shift can be easily observed in active sonar systems, particularly narrow-band ones, because the transmitter frequency is known, and the relative motion between sonar and object can be calculated. Sometimes the frequency of the radiated noise (a tonal) may also be known, in which case the same calculation can be done for passive sonar. For active systems the change in frequency is 0.69 Hz per knot per kHz and half this for passive systems as propagation is only one way. The shift corresponds to an increase in frequency for an approaching target.

Intensity Fluctuations

Though acoustic propagation modelling generally predicts a constant received sound level, in practice there are both temporal and spatial fluctuations. These may be due to both small and large scale environmental phenomena. These can include sound speed profile fine structure and frontal zones as well as internal waves. Because in general there are multiple propagation paths between a source and receiver, small phase changes in the interference pattern between these paths can lead to large fluctuations in sound intensity.

Non-linearity

In water, especially with air bubbles, the change in density due to a change in pressure is not exactly linearly proportional. As a consequence for a sinusoidal wave input additional harmonic and subharmonic frequencies are generated. When two sinusoidal waves are input, sum and difference frequencies are generated. The conversion process is greater at high source levels than small ones. Because of the non-linearity there is a dependence of sound speed on the pressure amplitude so that large changes travel faster than small ones. Thus a sinusoidal waveform gradually becomes a sawtooth one with a steep rise and a gradual tail. Use is made of this phenomenon in parametric sonar and theories have been developed to account for this, e.g. by Westerfield.

Measurements

Sound in water is measured using a hydrophone, which is the underwater equivalent of a microphone. A hydrophone measures pressure fluctuations, and these are usually converted to sound pressure level (SPL), which is a logarithmic measure of the mean square acoustic pressure.

Measurements are usually reported in one of three forms :-

- RMS acoustic pressure in micropascals (or dB re 1 µPa)

- RMS acoustic pressure in a specified bandwidth, usually octaves or thirds of octave (dB re 1 µPa)

- spectral density (mean square pressure per unit bandwidth) in micropascals-squared per hertz (dB re 1 $\mu Pa^2/Hz$)

The scale for acoustic pressure in water differs from that used for sound in air. In air the reference pressure is 20 µPa rather than 1 µPa. For the same numerical value of SPL, the intensity of a plane wave (power per unit area, proportional to mean square sound pressure divided by acoustic impedance) in air is about $20^2 \times 3600 = 1\,440\,000$ times higher than in water. Similarly, the intensity is about the same if the SPL is 61.6 dB higher in the water.

Sound Speed

Approximate values for fresh water and seawater, respectively, at atmospheric pressure are 1450 and 1500 m/s for the sound speed, and 1000 and 1030 kg/m³ for the density. The speed of sound in water increases with increasing pressure, temperature and salinity. The maximum speed in pure water under atmospheric pressure is attained at about 74 °C; sound travels slower in hotter water after that point; the maximum increases with pressure. On-line calculators can be found at Technical Guides - Speed of Sound in Sea-Water and Technical Guides - Speed of Sound in Pure Water.

Absorption

Many measurements have been made of sound absorption in lakes and the ocean.

Ambient Noise

Measurement of acoustic signals are possible if their amplitude exceeds a minimum threshold, determined partly by the signal processing used and partly by the level of background noise. Ambient noise is that part of the received noise that is independent of the source, receiver and platform characteristics. This it excludes reverberation and towing noise for example.

The background noise present in the ocean, or ambient noise, has many different sources and varies with location and frequency. At the lowest frequencies, from about 0.1 Hz to 10 Hz, ocean turbulence and microseisms are the primary contributors to the noise background. Typical noise spectrum levels decrease with increasing frequency from about 140 dB re 1 μPa^2/Hz at 1 Hz to about 30 dB re 1 μPa^2/Hz at 100 kHz. Distant ship traffic is one of the dominant noise sources in most areas for frequencies of around 100 Hz, while wind-induced surface noise is the main source between 1 kHz and 30 kHz. At very high frequencies, above 100 kHz, thermal noise of water molecules begins to dominate. The thermal noise spectral level at 100 kHz is 25 dB re 1 μPa^2/Hz. The spectral density of thermal noise increases by 20 dB per decade (approximately 6 dB per octave).

Transient sound sources also contribute to ambient noise. These can include intermittent geological activity, such as earthquakes and underwater volcanoes, rainfall on the surface, and biological activity. Biological sources include cetaceans (especially blue, fin and sperm whales), certain types of fish, and snapping shrimp.

Rain can produce high levels of ambient noise. However the numerical relationship between rain rate and ambient noise level is difficult to determine because measurement of rain rate is problematic at sea.

Reverberation

Many measurements have been made of sea surface, bottom and volume reverberation. Empirical models have sometimes been derived from these. A commonly used expression for the band 0.4 to 6.4 kHz is that by Chapman and Harris. It is found that a sinusoidal waveform is spread in frequency due to the surface motion. For bottom reverberation a Lambert's Law is found often to apply approximately. Volume reverberation is usually found to occur mainly in layers, which change depth with the time of day. The under-surface of ice can produce strong reverberation when it is rough.

Bottom Loss

Bottom loss has been measured as a function of grazing angle for many frequencies in various locations, for example those by the US Marine Geophysical Survey. The loss depends on the sound speed in the bottom (which is affected by gradients and layering) and by roughness. Graphs have been produced for the loss to be expected in particular circumstances. In shallow water bottom loss often has the dominant impact on long range propagation. At low frequencies sound can propagate through the sediment then back into the water.

Underwater Hearing

Comparison with Airborne Sound Levels

As with airborne sound, sound pressure level underwater is usually reported in units of decibels, but there are some important differences that make it difficult (and often inappropriate) to compare SPL in water with SPL in air. These differences include:

- difference in reference pressure: 1 μPa (one micropascal, or one millionth of a pascal) instead of 20 μPa.

- difference in interpretation: there are two schools of thought, one maintaining that pressures should be compared directly, and that the other that one should first convert to the intensity of an equivalent plane wave.

- difference in hearing sensitivity: any comparison with (A-weighted) sound in air needs to take into account the differences in hearing sensitivity, either of a human diver or other animal.

Hearing Sensitivity

The lowest audible SPL for a human diver with normal hearing is about 67 dB re 1 μPa, with greatest sensitivity occurring at frequencies around 1 kHz. This corresponds to a sound intensity 5.4 dB, or 3.5 times, higher than the threshold in air. Dolphins and other toothed whales are renowned for their acute hearing sensitivity, especially in the frequency range 5 to 50 kHz. Several species have hearing thresholds between 30 and 50 dB re 1 μPa in this frequency range. For example, the hearing threshold of the killer whale occurs at an RMS acoustic pressure of 0.02 mPa (and frequency 15 kHz), corresponding to an SPL threshold of 26 dB re 1 μPa. By comparison the most sensitive fish is the soldier fish, whose threshold is 0.32 mPa (50 dB re 1 μPa) at 1.3 kHz, whereas the lobster has a hearing threshold of 1.3 Pa at 70 Hz (122 dB re 1 μPa).

Safety Thresholds

High levels of underwater sound create a potential hazard to marine and amphibious animals as well as to human divers. Guidelines for exposure of human divers and marine mammals to underwater sound are reported by the SOLMAR project of the NATO Undersea Research Centre. Human divers exposed to SPL above 154 dB re 1 μPa in the frequency range 0.6 to 2.5 kHz are reported to experience changes in their heart rate or breathing frequency. Diver aversion to low frequency sound is dependent upon sound pressure level and center frequency.

Applications of Underwater Acoustics

Sonar

Sonar is the name given to the acoustic equivalent of radar. Pulses of sound are used to probe the sea, and the echoes are then processed to extract information about the sea, its boundaries and submerged objects. An alternative use, known as *passive sonar*, attempts to do the same by listening to the sounds radiated by underwater objects.

Underwater Communication

The need for underwater acoustic telemetry exists in applications such as data harvesting for environmental monitoring, communication with and between manned and unmanned underwater vehicles, transmission of diver speech, etc. A related application is underwater remote control, in which acoustic telemetry is used to remotely actuate a switch or trigger an event. A prominent example of underwater remote control are acoustic releases, devices that are used to return sea floor deployed instrument packages or other payloads to the surface per remote command at the end of a deployment. Acoustic communications form an active field of research with significant challenges to overcome, especially in horizontal, shallow-water channels. Compared with radio telecommunications, the available bandwidth is reduced by several orders of magnitude. Moreover, the low speed of sound causes multipath propagation to stretch over time delay intervals of tens or hundreds of milliseconds, as well as significant Doppler shifts and spreading. Often acoustic communication systems are not limited by noise, but by reverberation and time variability beyond the capability of receiver algorithms. The fidelity of underwater communication links can be greatly improved by the use of hydrophone arrays, which allow processing techniques such as adaptive beamforming and diversity combining.

Underwater Navigation and Tracking

Underwater navigation and tracking is a common requirement for exploration and work by divers, ROV, autonomous underwater vehicles (AUV), manned submersibles and submarines alike. Unlike most radio signals which are quickly absorbed, sound propagates far underwater and at a rate that can be precisely measured or estimated. It can thus be used to measure distances between a tracked target and one or multiple reference of *baseline stations* precisely, and triangulate the position of the target, sometimes with centimeter accuracy. Starting in the 1960s, this has given rise to underwater acoustic positioning systems which are now widely used.

Seismic Exploration

Seismic exploration involves the use of low frequency sound (< 100 Hz) to probe deep into the seabed. Despite the relatively poor resolution due to their long wavelength, low frequency sounds are preferred because high frequencies are heavily attenuated when they travel through the seabed. Sound sources used include airguns, vibroseis and explosives.

Weather and Climate Observation

Acoustic sensors can be used to monitor the sound made by wind and precipitation. For example, an acoustic rain gauge is described by Nystuen. Lightning strikes can also be detected.Acoustic thermometry of ocean climate (ATOC) uses low frequency sound to measure the global ocean temperature.

Oceanography

Large scale ocean features can be detected by acoustic tomography. Bottom characteristics can be measured by side-scan sonar and sub-bottom profiling.

Marine Biology

Due to its excellent propagation properties, underwater sound is used as a tool to aid the study of marine life, from microplankton to the blue whale. Echo sounders are often used to provide data on marine life abundance, distribution, and behavior information. Echo sounders, also referred to as hydroacoustics is also used for fish location, quantity, size, and biomass.

Acoustic telemetry is also used for monitoring fishes and marine wildlife. An acoustic transmitter is attached to the fish (sometimes internally) while an array of receivers listen to the information conveyed by the sound wave. This enables the researchers to track the movements of individuals in a small-medium scale.

Pistol shrimp create sonoluminescent cavitation bubbles that reach up to 5,000 K (4,700 °C)

Particle Physics

A neutrino is a fundamental particle that interacts very weakly with other matter. For this reason, it requires detection apparatus on a very large scale, and the ocean is sometimes used for this purpose. In particular, it is thought that ultra-high energy neutrinos in seawater can be detected acoustically.

Fisheries Acoustics

Fisheries acoustics includes a range of research and practical application topics using acoustical devices as sensors in aquatic environments. Acoustical techniques can be applied to sensing aquatic animals, zooplankton, and physical and biological habitat characteristics.

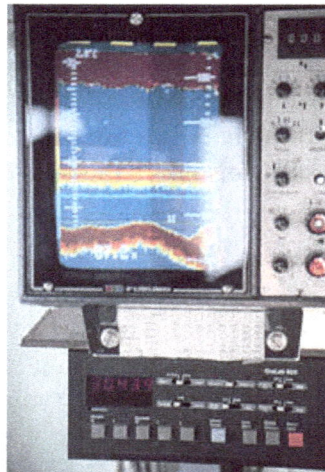

Fishfinder sonar

Basic Theory

Biomass estimation is a method of detecting and quantifying fish and other marine organisms using sonar technology. An acoustic transducer emits a brief, focused pulse of sound into the water. If the sound encounters objects that are of different density than the surrounding medium, such as fish, they reflect some sound back toward the source. These echoes provide information on fish size, location, and abundance. The basic components of the scientific echo sounder hardware function is to transmit the sound, receive, filter and amplify, record, and analyze the echoes. While there are many manufacturers of commercially available "fish-finders," quantitative analysis requires that measurements be made with calibrated echo sounder equipment, having high signal-to-noise ratios.

History

An extremely wide variety of fish taxa produce sound. Sound production behavior provides an opportunity to study various aspects of fish biology, such as spawning behavior and habitat selection, in a noninvasive manner. Passive acoustic methods can be an attractive alternative or supplement to traditional fisheries assessment techniques because they are noninvasive, can be conducted at low cost, and can cover a large study area at high spatial and temporal resolution.

Following the First World War, when sonar was first used for the detection of submarines, echo sounders began to find uses outside the military. The French explorer Rallier du Baty reported unexpected midwater echoes, which he attributed to fish schools, in 1927. In 1929, the Japanese scientist Kimura reported disruptions in a continuous acoustic beam by sea bream swimming in an aquaculture pond.

In the early 1930s, two commercial fishermen, Ronald Balls, an Englishman, and Reinert Bokn, a Norwegian, began independently experimenting with echosounders as a means to locate fish. Acoustic traces of sprat schools recorded by Bokn in Frafjord, Norway was the first echogram of fish to be published. In 1935, Norwegian scientist Oscar Sund reported observations of cod schools from the research vessel Johan Hjort, marking the first use of echosounding for fisheries research.

Sonar technologies developed rapidly during the Second World War, and military surplus equipment was adopted by commercial fishers and scientists soon after the end of hostilities. This period saw the first development of instruments designed specifically to detect fish. Large uncertainties persisted in the interpretation of acoustic surveys, however: calibration of instruments was irregular and imprecise, and the sound-scattering properties of fish and other organisms was poorly understood. Beginning in the 1970s and 80s, a series of practical and theoretical investigations began to overcome these limitations. Technological advances such as split-beam echosounders, digital signal processing, and electronic displays also appeared in this period.

At present, acoustic surveys are used in the assessment and management of many fisheries worldwide. Calibrated, split-beam echosounders are the standard equipment. Several acoustic frequencies are often used simultaneously, allowing some discrimination of different types of animals. Technological development continues, including research into multibeam, broadband, and parametric sonars.

Techniques

Fish Counting

When individual targets are spaced far enough apart that they can be distinguished from one another, it is straightforward to estimate the number of fish by counting the number of targets. This type of analysis is called echo counting, and was historically the first to be used for biomass estimation.

Echo Integration

If more than one target is located in the acoustic beam at the same depth, it is not usually possible to resolve them separately. This is often the case with schooling fish or aggregations of zooplankton. In these cases, echo integration is used to estimate biomass. Echo integration assumes that the total acoustic energy scattered by a group of targets is the sum of the energy scattered by each individual target. This assumption holds well in most cases. The total acoustic energy backscattered by the school or aggregation is integrated together, and this total is divided by the (previously determined) backscattering coefficient of a single animal, giving an estimate of the total number.

Instruments

Echosounders

The primary tool in fisheries acoustics is the scientific echosounder. This instrument operates on the same principles as a recreational or commercial fishfinder or echosounder, but is engineered for greater accuracy and precision, allowing quantitative biomass estimates to be made. In an echosounder, a transceiver generates a short pulse which is sent into the water by the transducer, an array of piezoelectric elements arranged to produce a focused beam of sound. In order to be used for quantitative work, the echosounder must be calibrated in the same configuration and environment in which it will be used; this is typically done by examining echoes from a metal sphere with known acoustic properties.

Early echosounders only transmitted a single beam of sound. Because of the acoustic beam pattern, identical targets at different azimuth angles will return different echo levels. If the beam pattern and angle to the target are known, this directivity can be compensated for. The need to determine the angle to a target led to the development of the twin-beam echosounder, which forms two acoustic beams, one inside the other. By comparing the phase difference of the same echo in the inner and outer beams, the angle off-axis can be estimated. In a further refinement of this concept, a split-beam echosounder divides the transducer face into four quadrants, allowing the location of targets in three dimensions. Single-frequency, split-beam echosounders are now the standard instrument of fisheries acoustics.

Multibeam Echosounders

Multibeam sonars project a fan-shaped set of sound beams outward into the water and record echoes in each beam. These have been widely used in bathymetric surveys, but have recently begun to find use in fisheries acoustics as well. Their major advantage is the addition of a second dimension to the narrow water column profile given by an echosounder. Multiple pings can thus be combined to give a three-dimensional picture of animal distributions.

Acoustic Cameras

Acoustic cameras are instruments that image a three-dimensional volume of water instantaneously. These typically use higher-frequency sound than traditional echosounders. This increases their resolution so that individual objects can be seen in detail, but means that their range is limited to tens of meters. They can be very useful for studying fish behavior in enclosed and/or murky bodies of water, for instance monitoring the passage of anadromous fish at dams

Platforms for Fisheries Acoustics

Fisheries acoustic research is conducted from a variety of platforms. The most common is a traditional research vessel, with the echosounders mounted on the ship's hull or in a drop keel. If the vessel does not have permanently installed echosounders, they may be deployed on a pole mount attached to the ship's side, or on a towed body or "towfish" pulled behind or alongside the vessel. Towed bodies are particularly useful for studies of deep-living fish, such as the orange roughy, which typically live below the range of an echosounder at the surface.

In addition to research vessels, acoustic data may be collected from a variety of "ships of opportunity" such as fishing vessels, ferries, and cargo ships. Ships of opportunity can offer low-cost data collection over large areas, though the lack of a true survey design may make analysis of these data difficult. In recent years, acoustic instruments have also been deployed on remotely operated vehicles and autonomous underwater vehicles, as well as at ocean observatories.

Target Strength Observations and Modelling

Target strength (TS) is a measurement of how well a fish, zooplankter, or other target scatters sound back towards the transducer. In general, larger animals have larger target strengths, though other factors, such as the presence or absence of a gas-filled swimbladder in fishes, may have a much larger effect. Target strength is of critical importance in fisheries acoustics, since it provides a link between acoustic backscatter and animal biomass. TS can be derived theoretically for simple targets such as spheres and cylinders, but in practice, it is usually measured empirically or calculated with numerical models.

Archaeoacoustics

Archaeoacoustics is the use of acoustical study as a methodological approach within archaeology. This may for example involve the study of the acoustics of archaeological sites, or the study of the acoustics of archaeological artefacts. Since many cultures explored through archaeology were focused on the oral and therefore the aural, it is becoming increasingly recognised that studying the sonic nature of parts of archaeology can enhance our understanding. This is an interdisciplinary field which includes areas such as archaeology, ethnomusicology, acoustics and digital modelling, and that is a part of the wider field of music archaeology. There is particular interest in prehistoric music.

Notable Work

Dr. Aaron Watson undertook work on the acoustics of numerous archaeological sites, including that of Stonehenge. He also investigated numerous chamber tombs and other stone circles. Archaeologist Paul Devereux's work has looked at ringing rocks, Avebury and various other subjects, and his book *Stone Age Soundtracks* provides a wide overview. Dr. Ian Cross of Cambridge University has explored lithoacoustics, the use of stones as musical instruments. Dr. Rupert Till of Huddersfield University has also explored Stonehenge's acoustics, along with Dr. Bruno Fazenda of Salford University. Dr. Damian Murphy of the University of York has studied measurement techniques in acoustic archaeology. Iegor Reznikoff and Michel Dauvois have studied the prehistoric painted caves of France, and found links between the artworks' positioning and acoustic effects. An AHRC project headed by Dr. Rupert Till of Huddersfield University, Professor Chris Scarre of Durham University and Dr. Bruno Fazenda of Salford University, studies similar relationships in the prehistoric painted caves in northern Spain. Steven Waller has also studied the links between rock art and sound. Panagiotis Karampatzakis and Vasilios Zafranas investigate the Acoustic Properties of Acheron Nekromantio, Aristoxenus acoustic vases, and the evolution of acoustics in the ancient greek and roman odea. Miriam Kolar and colleagues from Stanford University studied various spatial and perceptual attributes of Chavín de Huántar.

Networks

The International Study Group on Music Archaeology (ISGMA) includes archaeoacoustical work. It is a pool of researchers devoted to the field of music archaeology. The study group is hosted at the Orient Department of the German Archaeological Institute Berlin (DAI, Deutsches Archäologisches Institut, Orient-Abteilung) and the Department for Ethnomusicology at the Ethnological Museum Berlin (Ethnologisches Museum Berlin, SMB SPK, Abteilung Musikethnologie, Medien-Technik und Berliner Phonogramm-Archiv). It comprises research methods of musicological and anthropological disciplines, such as archaeology, organology, acoustics, music iconology, philology, ethnohistory, and ethnomusicology. The Acoustics and Music of British Prehistory Research Network was funded by the Arts and Humanities Research Council and Engineering and Physical Sciences Research Council, led by Rupert Till and Chris Scarre, as well as Professor Jian Kang of Sheffield University's Department of Architecture. It has a list of researchers working in the field, and links to many other relevant sites. An e-mail list has been discussing the subject since 2002 and was set up as a result of the First Pan-American/Iberian Meeting on Acoustics by Victor Reijs.

Past Interpretations Controversy

An early interpretation of the idea of archaeoacoustics was that it explored acoustic phenomena encoded in ancient artifacts. For instance, the idea that a pot or vase could be "read" like a gramophone record or phonograph cylinder for messages from the past, sounds encoded into the turning clay as the pot was thrown. There is little evidence to support such ideas, and there are few publications claiming that this is the case. In comparison, the more contemporary approach to the field now has many publications and a growing significance. This earlier approach was first raised in the

6 February 1969 issue of *New Scientist* magazine, where it was discussed in David E. H. Jones's light-hearted "Daedalus" column. He wrote:

[A] trowel, like any flat plate, must vibrate in response to sound: thus, drawn over the wet surface by the singing plasterer, it must emboss a gramophone-type recording of his song in the plaster. Once the surface is dry, it may be played back.

—Jones, 1982

Jones subsequently received a letter from one Richard G. Woodbridge III who claimed to have already been working on the idea and said that he had sent a paper on the subject to the journal *Nature*. The paper never appeared in *Nature*, but the August 1969 edition of the journal *Proceedings of the IEEE* printed a letter from Woodbridge entitled "Acoustic Recordings from Antiquity". In this communication, the author stated that he wished to call attention to the potential of what he called "Acoustic Archaeology" and to record some early experiments in the field. He then described his experiments with making clay pots and oil paintings from which sound could then be replayed, using a conventional record player cartridge connected directly to a set of headphones. He claimed to have extracted the hum of the potter's wheel from the grooves of a pot, and the word "blue" from an analysis of patch of blue color in a painting.

In 1993, archeology professor Paul Åström and acoustics professor Mendel Kleiner performed similar experiments in Gothenburg, and reported that they could recover some sounds.

An episode of *Mythbusters* explored the idea; Episode 62: Killer Cable Snaps, Pottery Record found that while *some* generic acoustic phenomena can be found on pottery, it is unlikely that any discernible sounds (like someone talking) could be recorded on the pots unless the ancient peoples had the technical knowledge to deliberately put the sounds on the artifacts.

As early as circa 1902, Charles Sanders Peirce wrote: "Give science only a hundred more centuries of increase in geometrical progression, and she may be expected to find that the sound waves of Aristotle's voice have somehow recorded themselves."

In Popular Culture

- An episode of Mysteryquest on History called Stonehenge featured Rupert Till and Bruno Fazenda conducting acoustic tests at Stonehenge and at the Maryhill Monument, a full-sized replica of Stonehenge in the USA.

- Gregory Benford's 1979 short story "Time Shards" concerns a researcher who recovers thousand-year-old sound from a piece of pottery thrown on a wheel and inscribed with a fine wire as it spun. The sound is then analyzed to reveal conversations between the potter and his assistant in Middle English.

- Rudy Rucker's 1981 short story "Buzz" includes a small section of audio recovered from ancient Egyptian pottery.

- A 2000 episode of *The X-Files*, "Hollywood A.D.", features "The Lazarus Bowl", a mythical piece of pottery reputed to have recorded on it the words that Jesus Christ spoke when he raised Lazarus from the dead.

- In the 1996 game *Amber: Journeys Beyond*, this phenomenon is referred to as "stone tape theory" and a key part of the game's plot.

- CSI: Crime Scene Investigation used this in 2005 episode *Committed*, where an inmate's conversation is partially recorded on a clay jar.

- In the first season episode of Fringe entitled "The Road Not Taken", an electron microscope is used to reproduce sounds captured on a partially melted window.

Physical Acoustics

Physical acoustics is the area of acoustics and physics that studies interactions of acoustic waves with a gaseous, liquid or solid medium on macro- and micro-levels. This relates to the interaction of sound with thermal waves in crystals (phonons), with light (photons), with electrons in metals and semiconductors (acousto-electric phenomena), with magnetic excitations in ferromagnetic crystals (magnons), etc. Some recently developed experimental techniques include photo-acoustics, acoustic microscopy and acoustic emission. A long standing interest is in acoustic and ultrasonic wave propagation and scattering in inhomogeneous materials, including composite materials and biological tissues.

There are two main classes of problems studied in physical acoustics. The first one concerns understanding how the physical properties of a medium (solid, liquid, or gas) influence the propagation of acoustic waves in this medium in order to use this knowledge for practical purposes. The second important class of problems studied in physical acoustics is to obtain the relevant information about a medium under consideration by measuring the properties of acoustic waves propagating through this medium.

References

- Templeton, Duncan (1993). Acoustics in the Built Environment: Advice for the Design Team. Architectural Press. ISBN 978-0750605380.

- Hamilton, M.F.; Blackstock, D.T. (1998). Nonlinear Acoustics. Academic Press. p. 55. ISBN 0-12-321860-8.

- David E.H., Jones (1982), The Inventions of Daedalus: A Compendium of Plausible .Schemes, W.H. Freeman & Company, ISBN 0-7167-1412-4

- Pierce, A.D. (1989), Acoustics: An Introduction to its Physical Principles and Applications, Acoustical Society of America, ISBN 0883186128

- Mason, W.P.; Thurston, R.N., eds. (1999), Physical Acoustics: Principles and Methods, 24, Academic Press, ISBN 012477945X

- Martignac F., Daroux A. , Baglinière J.L., Ombredanne D., Guilalrd J., 2015. The use of acoustic cameras in shallow waters: new hydroacoustic tools for monitoring migratory fish population. A review of DIDSON technology. Fish & Fisheries, 16 (3), 486–510. DOI: 10.1111/faf.12071

- M. Pourhomayoun, P. Dugan, M. Popescu, and C. Clark, "Bioacoustic Signal Classification Based on Continuous Region Features, Grid Masking Features and Artificial Neural Network," International Conference on Machine Learning (ICML), 2013.

- Wilson, Wayne D. (26 Jan 1959). "Speed of Sound in Distilled Water as a Function of Temperature and Pressure". J. Acoust. Soc. Am. 31 (8): 1067–1072. Bibcode:1959ASAJ...31.1067W. doi:10.1121/1.1907828. Retrieved 11 February 2012

- The Emulation of Nonlinear Time-Invariant Audio Systems with Memory by Means of Volterra Series, JAES Volume 60 Issue 12 pp. 984-996; December 2012.

- "Bioacoustics - the International Journal of Animal Sound and its Recording". Taylor & Francis. Retrieved 31 July 2012.

- V. F. Humphrey. "Non-linear propagation for medical imaging" (PDF). Department of Physics, University of Bath, Bath, UK. Retrieved 2008-11-10.

- Wells, P. N. T. (1999). "Ultrasonic imaging of the human body". Reports on Progress in Physics. 62 (5): 671. Bibcode:1999RPPh...62..671W. doi:10.1088/0034-4885/62/5/201.

- NATO Undersea Research Centre Human Diver and Marine Mammal Risk Mitigation Rules and Procedures, NURC Special Publication NURC-SP-2006-008, September 2006

- Fothergill DM, Sims JR, Curley MD (2001). "Recreational scuba divers' aversion to low-frequency underwater sound". Undersea Hyperb Med. 28 (1): 9–18. PMID 11732884. Retrieved 2009-03-31.

Fundamental Concepts of Acoustics

To gain a meaningful understanding of acoustics, it is imperative to understand its basic concepts. This chapter discusses some of the important concepts like acoustic wave equation, fluid, P-wave and diffraction.

Acoustic Wave Equation

In physics, the acoustic wave equation governs the propagation of acoustic waves through a material medium. The form of the equation is a second order partial differential equation. The equation describes the evolution of acoustic pressure p or particle velocity u as a function of position x and time . A simplified form of the equation describes acoustic waves in only one spatial dimension, while a more general form describes waves in three dimensions.

The acoustic wave equation was an important point of reference in the development of the electromagnetic wave equation in Kelvin's master class at Johns Hopkins University.

For lossy media, more intricate models need to be applied in order to take into account frequency-dependent attenuation and phase speed. Such models include acoustic wave equations that incorporate fractional derivative terms.

In One Dimension

Equation

Richard Feynman derives the wave equation that describes the behaviour of sound in matter in one dimension (position x) as:

$$\frac{\partial^2 p}{\partial x^2} - \frac{1}{c^2}\frac{\partial^2 p}{\partial t^2} = 0$$

where p is the acoustic pressure (the local deviation from the ambient pressure), and where c is the speed of sound.

Solution

Provided that the speed c is a constant, not dependent on frequency (the dispersionless case), then the most general solution is

$$p = f(ct - x) + g(ct + x)$$

where f and g are any two twice-differentiable functions. This may be pictured as the superposition of two waveforms of arbitrary profile, one (f) travelling up the x-axis and the other (g) down the x-axis at the speed c. The particular case of a sinusoidal wave travelling in one direction is obtained by choosing either f or g to be a sinusoid, and the other to be zero, giving

$$p = p_0 \sin(\omega t \mp kx).$$

where ω is the angular frequency of the wave and k is its wave number.

Derivation

The wave equation can be developed from the linearized one-dimensional continuity equation, the linearized one-dimensional force equation and the equation of state.

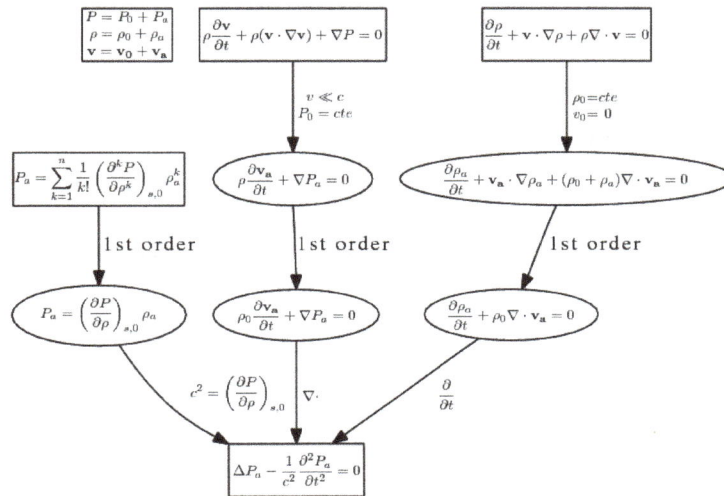

Derivation of the acoustic wave equation

The Equation of State (Ideal Gas Law)

$$PV = nRT$$

In an adiabatic process, pressure P as a function of density ρ can be linearized to

$$P = C\rho$$

where C is some constant. Breaking the pressure and density into their mean and total components and noting that $C = \dfrac{\partial P}{\partial \rho}$:

$$P - P_0 = \left(\frac{\partial P}{\partial \rho}\right)(\rho - \rho_0).$$

The adiabatic bulk modulus for a fluid is defined as

$$B = \rho_0 \left(\frac{\partial P}{\partial \rho}\right)_{adiabatic}$$

which gives the result

$$P - P_0 = B\frac{\rho - \rho_0}{\rho_0}.$$

Condensation, s, is defined as the change in density for a given ambient fluid density.

$$s = \frac{\rho - \rho_0}{\rho_0}$$

The linearized equation of state becomes

$p = Bs$ where p is the acoustic pressure ($P - P_0$).

The continuity equation (conservation of mass) in one dimension is

$$\frac{\partial \rho}{\partial t} + \frac{\partial}{\partial x}(\rho u) = 0 .$$

Where u is the flow velocity of the fluid. Again the equation must be linearized and the variables split into mean and variable components.

$$\frac{\partial}{\partial t}(\rho_0 + \rho_0 s) + \frac{\partial}{\partial x}(\rho_0 u + \rho_0 su) = 0$$

Rearranging and noting that ambient density does not change with neither time nor position and that the condensation multiplied by the velocity is a very small number:

$$\frac{\partial s}{\partial t} + \frac{\partial}{\partial x}u = 0$$

Euler's Force equation (conservation of momentum) is the last needed component. In one dimension the equation is:

$$\rho \frac{Du}{Dt} + \frac{\partial P}{\partial x} = 0 ,$$

where D/Dt represents the convective, substantial or material derivative, which is the derivative at a point moving with medium rather than at a fixed point.

Linearizing the variables:

$$(\rho_0 + \rho_0 s)\left(\frac{\partial}{\partial t} + u\frac{\partial}{\partial x}\right)u + \frac{\partial}{\partial x}(P_0 + p) = 0 .$$

Rearranging and neglecting small terms, the resultant equation becomes the linearized one-dimensional Euler Equation:

$$\rho_0 \frac{\partial u}{\partial t} + \frac{\partial p}{\partial x} = 0 .$$

Taking the time derivative of the continuity equation and the spatial derivative of the force equation results in:

$$\frac{\partial^2 s}{\partial t^2} + \frac{\partial^2 u}{\partial x \partial t} = 0$$

$$\rho_0 \frac{\partial^2 u}{\partial x \partial t} + \frac{\partial^2 p}{\partial x^2} = 0$$

Multiplying the first by ρ_0, subtracting the two, and substituting the linearized equation of state,

$$-\frac{\rho_0}{B}\frac{\partial^2 p}{\partial t^2} + \frac{\partial^2 p}{\partial x^2} = 0 .$$

The final result is

$$\frac{\partial^2 p}{\partial x^2} - \frac{1}{c^2}\frac{\partial^2 p}{\partial t^2} = 0$$

where $c = \sqrt{\dfrac{B}{\rho_0}}$ is the speed of propagation.

In Three Dimensions

Equation

Feynman provided a derivation of the wave equation that describes the behaviour of sound in matter in three dimensions as:

$$\nabla^2 p - \frac{1}{c^2}\frac{\partial^2 p}{\partial t^2} = 0$$

where ∇^2 is the Laplace operator, p is the acoustic pressure (the local deviation from the ambient pressure), and where c is the speed of sound.

A similar looking wave equation but for the vector field particle velocity is given by

$$\nabla^2 \mathbf{u} - \frac{1}{c^2}\frac{\partial^2 \mathbf{u}}{\partial t^2} = 0 .$$

In some situations, it is more convenient to solve the wave equation for an abstract scalar field velocity potential which has the form

$$\nabla^2 \Phi - \frac{1}{c^2}\frac{\partial^2 \Phi}{\partial t^2} = 0$$

and then derive the physical quantities particle velocity and acoustic pressure by the equations (or definition, in the case of particle velocity):

$$\mathbf{u} = \nabla \Phi ,$$

$$p = -\rho \frac{\partial}{\partial t}\Phi .$$

Solution

The following solutions are obtained by separation of variables in different coordinate systems. They are phasor solutions, that is they have an implicit time-dependence factor of $e^{i\omega t}$ where $\omega = 2\pi f$ is the angular frequency. The explicit time dependence is given by

$$p(r,t,k) = \text{Real}\left[p(r,k)e^{i\omega t} \right]$$

Here $k = \omega / c$ is the wave number.

Cartesian Coordinates

$$p(r,k) = Ae^{\pm ikr} .$$

Cylindrical Coordinates

$$p(r,k) = AH_0^{(1)}(kr) + BH_0^{(2)}(kr) .$$

where the asymptotic approximations to the Hankel functions, when $kr \to \infty$, are

$$H_0^{(1)}(kr) \simeq \sqrt{\frac{2}{\pi kr}} e^{i(kr - \pi/4)}$$

$$H_0^{(2)}(kr) \simeq \sqrt{\frac{2}{\pi kr}} e^{-i(kr - \pi/4)} \, .$$

Spherical Coordinates

$$p(r,k) = \frac{A}{r} e^{\pm ikr} \, .$$

Depending on the chosen Fourier convention, one of these represents an outward travelling wave and the other an unphysical inward travelling wave. The inward travelling solution wave is only unphysical because of the singularity that occurs at r=0; inward travelling waves do exist.

Fluid

In physics, a fluid is a substance that continually deforms (flows) under an applied shear stress. Fluids are a subset of the phases of matter and include liquids, gases, plasmas and, to some extent, plastic solids. Fluids are substances that have zero shear modulus or, in simpler terms, a fluid is a substance which cannot resist any shear force applied to it.

Although the term "fluid" includes both the liquid and gas phases, in common usage, "fluid" is often used as a synonym for "liquid", with no implication that gas could also be present. For example, "brake fluid" is hydraulic oil and will not perform its required incompressible function if there is gas in it. This colloquial usage of the term is also common in medicine and in nutrition ("take plenty of fluids").

Liquids form a free surface (that is, a surface not created by the container) while gases do not. The distinction between solids and fluid is not entirely obvious. The distinction is made by evaluating the viscosity of the substance. Silly Putty can be considered to behave like a solid or a fluid, depending on the time period over which it is observed. It is best described as a viscoelastic fluid. There are many examples of substances proving difficult to classify. A particularly interesting one is pitch, as demonstrated in the pitch drop experiment currently running at the University of Queensland.

Physics

Fluids display properties such as:

- not resisting deformation, or resisting it only slightly (viscosity), and

- the ability to flow (also described as the ability to take on the shape of the container).This also means that all liquids have the property of fluidity.

These properties are typically a function of their inability to support a shear stress in static equilibrium.

Solids can be subjected to shear stresses, and to normal stresses—both compressive and tensile. In contrast, ideal fluids can only be subjected to normal, compressive stress which is called pressure. Real fluids display viscosity and so are capable of being subjected to low levels of shear stress.

Modelling

In a solid, shear stress is a function of strain, but in a fluid, shear stress is a function of strain rate. A consequence of this behavior is Pascal's law which describes the role of pressure in characterizing a fluid's state.

Depending on the relationship between shear stress, and the rate of strain and its derivatives, fluids can be characterized as one of the following:

- Newtonian fluids: where stress is directly proportional to rate of strain

- Non-Newtonian fluids: where stress is not proportional to rate of strain, its higher powers and derivatives.

The behavior of fluids can be described by the Navier–Stokes equations—a set of partial differential equations which are based on:

- continuity (conservation of mass),

- conservation of linear momentum,

- conservation of angular momentum,

- conservation of energy.

The study of fluids is fluid mechanics, which is subdivided into fluid dynamics and fluid statics depending on whether the fluid is in motion.

P-Wave

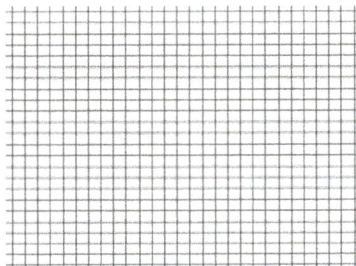

Plane P-wave

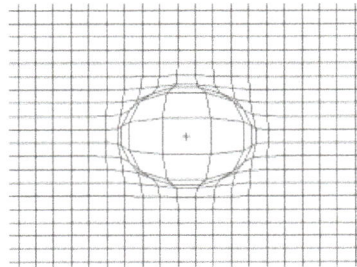

Representation of the propagation of a P-wave on a 2D grid (empirical shape)

P-waves are a type of body wave, called seismic waves in seismology, that travel through a continuum and are the first waves from an earthquake to arrive at a seismograph. The continuum is made up of gases (as sound waves), liquids, or solids, including the Earth. P-waves can be produced by earthquakes and recorded by seismographs. The name P-wave can stand for either pressure wave as it is formed from alternating compressions and rarefactions or primary wave, as it has the highest velocity and is therefore the first wave to be recorded.

In isotropic and homogeneous solids, the mode of propagation of a P-wave is always longitudinal; thus, the particles in the solid vibrate along the axis of propagation (the direction of motion) of the wave energy.

Velocity

The velocity of P-waves in a homogeneous isotropic medium is given by

$$v_p = \sqrt{\frac{K + \frac{4}{3}\mu}{\rho}} = \sqrt{\frac{\lambda + 2\mu}{\rho}}$$

where K is the bulk modulus (the modulus of incompressibility), μ is the shear modulus (modulus of rigidity, sometimes denoted as G and also called the second Lamé parameter), ρ is the density of the material through which the wave propagates, and λ is the first Lamé parameter.

Of these, density shows the least variation, so the velocity is mostly *controlled* by K and μ.

The elastic moduli P-wave modulus, M, is defined so that $M = K + 4\mu/3$ and thereby

$$v_p = \sqrt{M/\rho}$$

Typical values for P-wave velocity in earthquakes are in the range 5 to 8 km/s. The precise speed varies according to the region of the Earth's interior, from less than 6 km/s in the Earth's crust to 13 km/s through the core.

Velocity of Common Rock Types		
Rocktype	Velocity [m/s]	Velocity [ft/s]
Unconsolidated Sandstone	4600 - 5200	15000 - 17000
Consolidated Sandstone	5800	19000
Shale	1800 - 4900	6000 -16000
Limestone	5800 - 6400	19000 - 21000
Dolomite	6400 - 7300	21000 - 24000
Anhydrite	6100	20000
Granite	5800 - 6100	19000 - 20000
Gabbro	7200	23600

Geologist Francis Birch discovered a relationship between the velocity of P waves and the density of the material the waves are traveling in:

$$V_p = a(\bar{M}) + b\rho$$

which later became known as Birch's law.

Seismic Waves in The Earth

Primary and secondary waves are body waves that travel within the Earth. The motion and behavior of both *P*-type and *S*-type in the Earth are monitored to probe the interior structure of the Earth. Discontinuities in velocity as a function of depth are indicative of changes in phase or composition. Differences in arrival times of waves originating in a seismic event like an earthquake as a result of waves taking different paths allow mapping of the Earth's inner structure.

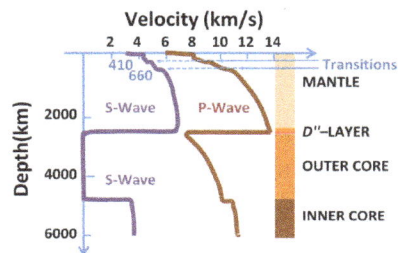

Velocity of seismic waves in the Earth versus depth. The negligible *S*-wave velocity in the outer core occurs because it is liquid, while in the solid inner core the *S*-wave velocity is non-zero.

P-Wave Shadow Zone

Almost all the information available on the structure of the Earth's deep interior is derived from observations of the travel times, reflections, refractions and phase transitions of seismic body waves, or normal modes. P-waves travel through the fluid layers of the Earth's interior, and yet they are refracted slightly when they pass through the transition between the semisolid mantle and the liquid outer core. As a result, there is a P-wave "shadow zone" between 103° and 142° from the earthquake's focus, where the initial P-waves are not registered on seismometers. In contrast, S-waves do not travel through liquids, rather, they are attenuated.

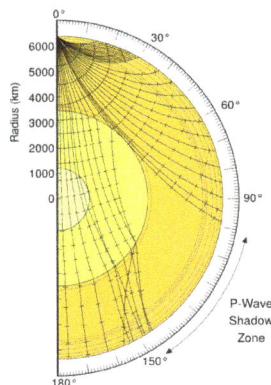

P-wave shadow zone (from USGS)

As An Earthquake Warning

Earthquake advance warning is possible by detecting the non-destructive primary waves that travel more quickly through the Earth's crust than do the destructive secondary and Rayleigh waves, in the same way that lightning flashes reach our eyes before we hear the thunder during a storm. The amount of advance warning depends on the delay between the arrival of the P-wave and other destructive waves, generally on the order of seconds up to about 60–90 seconds for deep, distant, large quakes such as Tokyo would have received before the 2011 Tohoku earthquake. The effectiveness of advance warning depends on accurate detection of the P-waves and rejection of ground vibrations caused by local activity (such as trucks or construction) as otherwise false-positive warnings will result. Earthquake Early Warning systems can be automated to allow for immediate safety actions such as issuing alerts, stopping elevators at the nearest floors or switching gas utilities off.

Acoustic Holography

Acoustic holography is a method that is used to estimate the sound field near a source by measuring acoustic parameters away from the source via an array of pressure and/or particle velocity transducers. Measuring techniques included within acoustic holography are becoming increasingly popular in various fields, most notably those of transportation, vehicle and aircraft design, and noise, vibration, and harshness (NVH). The general idea of acoustic holography has led to different versions such as near-field acoustic holography (NAH) and statistically optimal near-field acoustic holography (SONAH). For audio rendition, the wave field synthesis is the most related procedure.

Diffraction

Diffraction refers to various phenomena which occur when a wave encounters an obstacle or a slit. It is defined as the bending of light around the corners of an obstacle or aperture into the region of geometrical shadow of the obstacle. In classical physics, the diffraction phenomenon is described as the interference of waves according to the Huygens–Fresnel principle. These characteristic behaviors are exhibited when a wave encounters an obstacle or a slit that is comparable in size to its wavelength. Similar effects occur when a light wave travels through a medium with a varying refractive index, or when a sound wave travels through a medium with varying acoustic impedance. Diffraction occurs with all waves, including sound waves, water waves, and electromagnetic waves such as visible light, X-rays and radio waves.

Since physical objects have wave-like properties (at the atomic level), diffraction also occurs with matter and can be studied according to the principles of quantum mechanics. Italian scientist Francesco Maria Grimaldi coined the word "diffraction" and was the first to record accurate observations of the phenomenon in 1660.

While diffraction occurs whenever propagating waves encounter such changes, its effects are generally most pronounced for waves whose wavelength is roughly comparable to the dimensions of the diffracting object or slit. If the obstructing object provides multiple, closely spaced openings, a complex pattern of varying intensity can result. This is due to the addition, or interference, of different parts of a wave that travels to the observer by different paths, where different path lengths result in different phases. The formalism of diffraction can also describe the way in which waves of finite extent propagate in free space. For example, the expanding profile of a laser beam, the beam shape of a radar antenna and the field of view of an ultrasonic transducer can all be analyzed using diffraction equations.

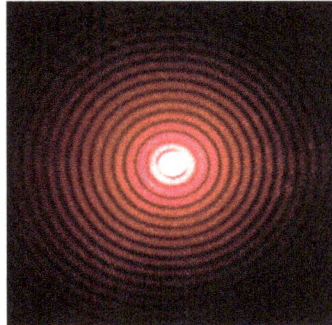

Diffraction pattern of red laser beam made on a plate after passing a small circular hole in another plate

Examples

The effects of diffraction are often seen in everyday life. The most striking examples of diffraction are those that involve light; for example, the closely spaced tracks on a CD or DVD act as a diffraction grating to form the familiar rainbow pattern seen when looking at a disk. This principle can be extended to engineer a grating with a structure such that it will produce any diffraction pattern desired; the hologram on a credit card is an example. Diffraction in the atmosphere by small particles can cause a bright ring to be visible around a bright light source like the sun or the moon. A shadow of a solid object, using light from a compact source, shows small fringes near its edges. The speckle pattern which is observed when laser light falls on an optically rough surface is also a diffraction phenomenon. When deli meat appears to be iridescent, that is diffraction off the meat fibers. All these effects are a consequence of the fact that light propagates as a wave.

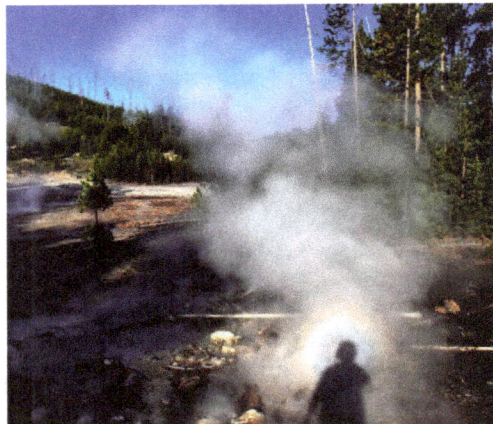

Solar glory at the steam from hot springs. A glory is an optical phenomenon produced by light backscattered (a combination of diffraction, reflection and refraction) towards its source by a cloud of uniformly sized water droplets.

Diffraction can occur with any kind of wave. Ocean waves diffract around jetties and other obstacles. Sound waves can diffract around objects, which is why one can still hear someone calling even when hiding behind a tree. Diffraction can also be a concern in some technical applications; it sets a fundamental limit to the resolution of a camera, telescope, or microscope.

History

The law of diffraction of light, which would much later come to be known as Snell's law was first accurately described by the scientist Ibn Sahl at the Baghdad court in 984. In the manuscript *On Burning Mirrors and Lenses*, Sahl used the law to derive lens shapes that focus light with no geometric aberrations.

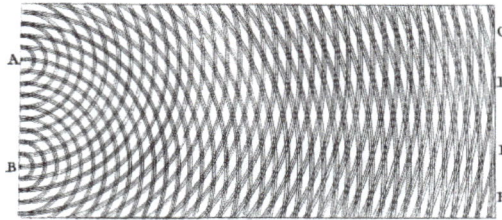

Thomas Young's sketch of two-slit diffraction, which he presented to the Royal Society in 1803.

The effects of diffraction of light were first carefully observed and characterized by Francesco Maria Grimaldi, who also coined the term *diffraction*, from the Latin *diffringere*, 'to break into pieces', referring to light breaking up into different directions. The results of Grimaldi's observations were published posthumously in 1665. Isaac Newton studied these effects and attributed them to *inflexion* of light rays. James Gregory (1638–1675) observed the diffraction patterns caused by a bird feather, which was effectively the first diffraction grating to be discovered. Thomas Young performed a celebrated experiment in 1803 demonstrating interference from two closely spaced slits. Explaining his results by interference of the waves emanating from the two different slits, he deduced that light must propagate as waves. Augustin-Jean Fresnel did more definitive studies and calculations of diffraction, made public in 1815 and 1818, and thereby gave great support to the wave theory of light that had been advanced by Christiaan Huygens and reinvigorated by Young, against Newton's particle theory.

Mechanism

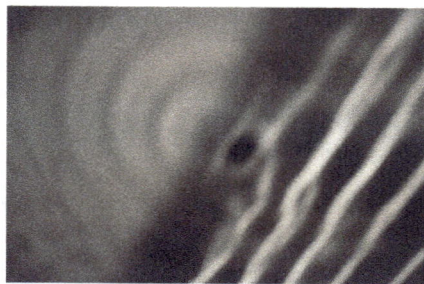

Photograph of single-slit diffraction in a circular ripple tank

Diffraction arises because of the way in which waves propagate; this is described by the Huygens–Fresnel principle and the principle of superposition of waves. The propagation of a wave can be visualized by considering every particle of the transmitted medium on a wavefront as a point source for a secondary spherical wave. The wave displacement at any subsequent point is the sum of these secondary waves. When waves are added together, their sum is determined by the relative phases as well as the amplitudes of the individual waves so that the summed amplitude of the waves can have any value between zero and the sum of the individual amplitudes. Hence, diffraction patterns usually have a series of maxima and minima.

There are various analytical models which allow the diffracted field to be calculated, including the Kirchhoff-Fresnel diffraction equation which is derived from wave equation, the Fraunhofer diffraction approximation of the Kirchhoff equation which applies to the far field and the Fresnel diffraction approximation which applies to the near field. Most configurations cannot be solved analytically, but can yield numerical solutions through finite element and boundary element methods.

It is possible to obtain a qualitative understanding of many diffraction phenomena by considering how the relative phases of the individual secondary wave sources vary, and in particular, the conditions in which the phase difference equals half a cycle in which case waves will cancel one another out.

The simplest descriptions of diffraction are those in which the situation can be reduced to a two-dimensional problem. For water waves, this is already the case; water waves propagate only on the surface of the water. For light, we can often neglect one direction if the diffracting object extends in that direction over a distance far greater than the wavelength. In the case of light shining through small circular holes we will have to take into account the full three-dimensional nature of the problem.

Diffraction of Light

Some examples of diffraction of light are considered below.

Single-slit Diffraction

Numerical approximation of diffraction pattern from a slit of width equal to wavelength of an incident plane wave in 3D spectrum visualization

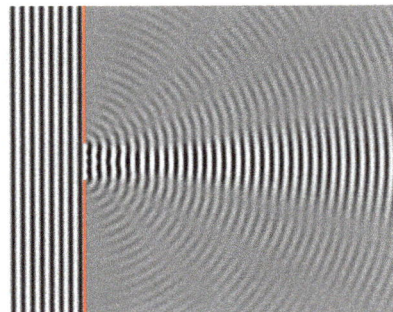

Numerical approximation of diffraction pattern from a slit of width four wavelengths with an incident plane wave. The main central beam, nulls, and phase reversals are apparent.

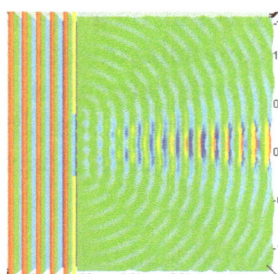

Numerical approximation of
diffraction pattern from a slit of width
equal to five times the wavelength of
an incident plane wave in 3D spectrum
visualization

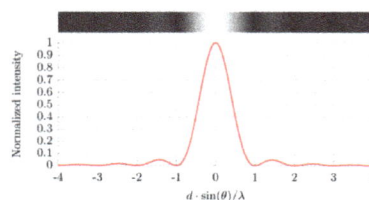

Graph and image of single-slit
diffraction.

A long slit of infinitesimal width which is illuminated by light diffracts the light into a series of circular waves and the wavefront which emerges from the slit is a cylindrical wave of uniform intensity.

A slit which is wider than a wavelength produces interference effects in the space downstream of the slit. These can be explained by assuming that the slit behaves as though it has a large number of point sources spaced evenly across the width of the slit. The analysis of this system is simplified if we consider light of a single wavelength. If the incident light is coherent, these sources all have the same phase. Light incident at a given point in the space downstream of the slit is made up of contributions from each of these point sources and if the relative phases of these contributions vary by 2π or more, we may expect to find minima and maxima in the diffracted light. Such phase differences are caused by differences in the path lengths over which contributing rays reach the point from the slit.

We can find the angle at which a first minimum is obtained in the diffracted light by the following reasoning. The light from a source located at the top edge of the slit interferes destructively with a source located at the middle of the slit, when the path difference between them is equal to $\lambda/2$. Similarly, the source just below the top of the slit will interfere destructively with the source located just below the middle of the slit at the same angle. We can continue this reasoning along the entire height of the slit to conclude that the condition for destructive interference for the entire slit is the same as the condition for destructive interference between two narrow slits a distance apart that is half the width of the slit. The path difference is given by $\dfrac{d \sin(\theta)}{2}$ so that the minimum intensity occurs at an angle θ_{min} given by

$$d \sin \theta_{\text{min}} = \lambda$$

where

- d is the width of the slit,

- θ_{min} is the angle of incidence at which the minimum intensity occurs, and

- λ is the wavelength of the light

A similar argument can be used to show that if we imagine the slit to be divided into four, six, eight parts, etc., minima are obtained at angles θ_n given by

$$d\sin\theta_n = n\lambda$$

where

- n is an integer other than zero.

There is no such simple argument to enable us to find the maxima of the diffraction pattern. The intensity profile can be calculated using the Fraunhofer diffraction equation as

$$I(\theta) = I_0 \,\mathrm{sinc}^2\left(\frac{d\pi}{\lambda}\sin\theta\right)$$

where

- $I(\theta)$ is the intensity at a given angle,
- I_0 is the original intensity, and
- the unnormalized sinc function above is given by $\mathrm{sinc}(x) = \dfrac{\sin x}{x}$ if $x \neq 0$, , and $\mathrm{sinc}(0) = 1$

This analysis applies only to the far field, that is, at a distance much larger than the width of the slit.

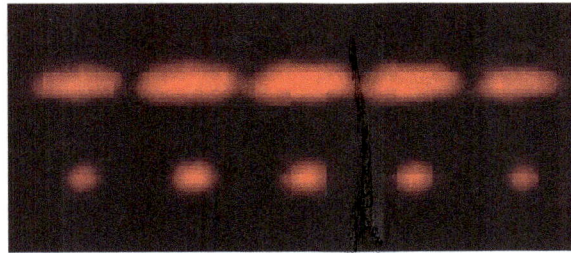

2-slit (top) and 5-slit diffraction of red laser light

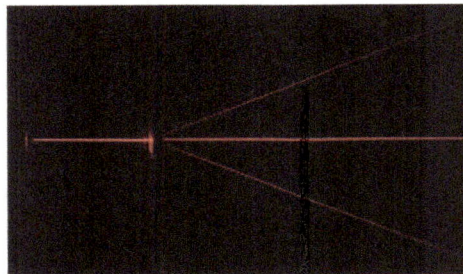

Diffraction of a red laser using a diffraction grating.

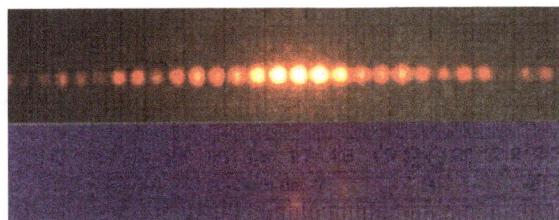

A diffraction pattern of a 633 nm laser through a grid of 150 slits

Diffraction Grating

A diffraction grating is an optical component with a regular pattern. The form of the light diffracted by a grating depends on the structure of the elements and the number of elements present, but all gratings have intensity maxima at angles θ_m which are given by the grating equation

$$d\left(\sin\theta_m + \sin\theta_i\right) = m\lambda.$$

where

- θ_i is the angle at which the light is incident,

- d is the separation of grating elements, and

- m is an integer which can be positive or negative.

The light diffracted by a grating is found by summing the light diffracted from each of the elements, and is essentially a convolution of diffraction and interference patterns.

The figure shows the light diffracted by 2-element and 5-element gratings where the grating spacings are the same; it can be seen that the maxima are in the same position, but the detailed structures of the intensities are different.

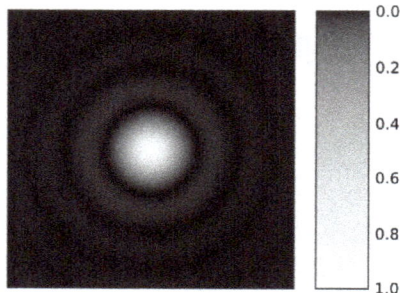

A computer-generated image of an Airy disk.

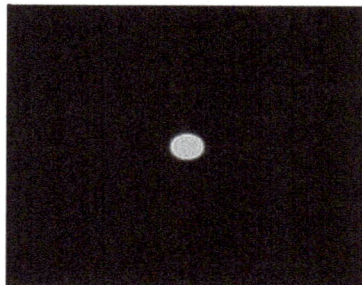

Computer generated light diffraction pattern from a circular aperture of diameter 0.5 micrometre at a wavelength of 0.6 micrometre (red-light) at distances of 0.1 cm – 1 cm in steps of 0.1 cm. One can see the image moving from the Fresnel region into the Fraunhofer region where the Airy pattern is seen.

Circular Aperture

The far-field diffraction of a plane wave incident on a circular aperture is often referred to as the Airy Disk. The variation in intensity with angle is given by

$$I(\theta) = I_0 \left(\frac{2J_1(ka\sin\theta)}{ka\sin\theta} \right)^2 ,$$

where a is the radius of the circular aperture, k is equal to $2\pi/\lambda$ and J_1 is a Bessel function. The smaller the aperture, the larger the spot size at a given distance, and the greater the divergence of the diffracted beams.

General Aperture

The wave that emerges from a point source has amplitude ψ at location r that is given by the solution of the frequency domain wave equation for a point source (The Helmholtz Equation),

$$\nabla^2\psi + k^2\psi = \delta(\mathbf{r})$$

where $\delta(\mathbf{r})$ is the 3-dimensional delta function. The delta function has only radial dependence, so the Laplace operator (a.k.a. scalar Laplacian) in the spherical coordinate system simplifies to:

$$\nabla^2\psi = \frac{1}{r}\frac{\partial^2}{\partial r^2}(r\psi)$$

By direct substitution, the solution to this equation can be readily shown to be the scalar Green's function, which in the spherical coordinate system (and using the physics time convention $e^{-i\omega t}$) is:

$$\psi(r) = \frac{e^{ikr}}{4\pi r}$$

This solution assumes that the delta function source is located at the origin. If the source is located at an arbitrary source point, denoted by the vector $\mathbf{r}'$ and the field point is located at the point $\mathbf{r}$, then we may represent the scalar Green's function (for arbitrary source location) as:

$$\psi(\mathbf{r}\,|\,\mathbf{r}') = \frac{e^{ik|\mathbf{r}-\mathbf{r}'|}}{4\pi\,|\,\mathbf{r}-\mathbf{r}'\,|}$$

Therefore, if an electric field, $E_{inc}(x,y)$ is incident on the aperture, the field produced by this aperture distribution is given by the surface integral:

$$\Psi(r) \propto \iint\limits_{aperture} E_{inc}(x',y')\frac{e^{ik|\mathbf{r}-\mathbf{r}'|}}{4\pi\,|\,\mathbf{r}-\mathbf{r}'\,|}dx'dy',$$

where the source point in the aperture is given by the vector

$$\mathbf{r}' = x'\hat{\mathbf{x}} + y'\hat{\mathbf{y}}$$

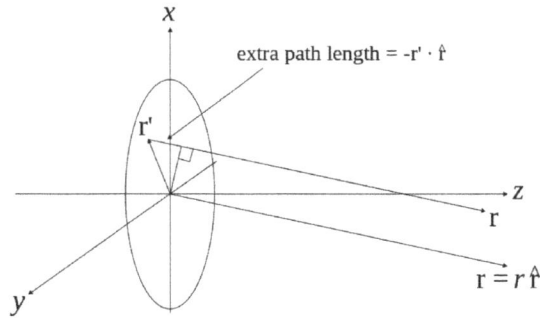

On The Calculation of Fraunhofer Region Fields

In the far field, wherein the parallel rays approximation can be employed, the Green's function,

$$\psi(\mathbf{r}\,|\,\mathbf{r}') = \frac{e^{ik|\mathbf{r}-\mathbf{r}'|}}{4\pi\,|\mathbf{r}-\mathbf{r}'|}$$

simplifies to

$$\psi(\mathbf{r}\,|\,\mathbf{r}') = \frac{e^{ikr}}{4\pi r}\,e^{-ik(\mathbf{r}'\cdot\hat{\mathbf{r}})}$$

as can be seen in the figure.

The expression for the far-zone (Fraunhofer region) field becomes

$$\Psi(r) \propto \frac{e^{ikr}}{4\pi r}\iint\limits_{\text{aperture}} E_{\text{inc}}(x',y')e^{-ik(\mathbf{r}'\cdot\hat{\mathbf{r}})}\,dx'dy',$$

Now, since

$$\mathbf{r}' = x'\hat{\mathbf{x}} + y'\hat{\mathbf{y}}$$

and

$$\hat{\mathbf{r}} = \sin\theta\cos\phi\hat{\mathbf{x}} + \sin\theta\sin\phi\,\hat{\mathbf{y}} + \cos\theta\hat{\mathbf{z}}$$

the expression for the Fraunhofer region field from a planar aperture now becomes,

$$\Psi(r) \propto \frac{e^{ikr}}{4\pi r}\iint\limits_{\text{aperture}} E_{\text{inc}}(x',y')e^{-ik\sin\theta(\cos\phi x'+\sin\phi y')}\,dx'dy'$$

Letting,

$$k_x = k\sin\theta\cos\phi$$

and

$$k_y = k \sin \theta \sin \phi$$

the Fraunhofer region field of the planar aperture assumes the form of a Fourier transform

$$\Psi(r) \propto \frac{e^{ikr}}{4\pi r} \iint\limits_{\text{aperture}} E_{\text{inc}}(x', y') e^{-i(k_x x' + k_y y')} \, dx' dy',$$

In the far-field / Fraunhofer region, this becomes the spatial Fourier transform of the aperture distribution. Huygens' principle when applied to an aperture simply says that the far-field diffraction pattern is the spatial Fourier transform of the aperture shape, and this is a direct by-product of using the parallel-rays approximation, which is identical to doing a plane wave decomposition of the aperture plane fields.

Propagation of A Laser Beam

The way in which the beam profile of a laser beam changes as it propagates is determined by diffraction. When the entire emitted beam has a planar, spatially coherent wave front, it approximates Gaussian beam profile and has the lowest divergence for a given diameter. The smaller the output beam, the quicker it diverges. It is possible to reduce the divergence of a laser beam by first expanding it with one convex lens, and then collimating it with a second convex lens whose focal point is coincident with that of the first lens. The resulting beam has a larger diameter, and hence a lower divergence. Divergence of a laser beam may be reduced below the diffraction of a Gaussian beam or even reversed to convergence if the refractive index of the propagation media increases with the light intensity. This may result in a self-focusing effect.

When the wave front of the emitted beam has perturbations, only the transverse coherence length (where the wave front perturbation is less than 1/4 of the wavelength) should be considered as a Gaussian beam diameter when determining the divergence of the laser beam. If the transverse coherence length in the vertical direction is higher than in horizontal, the laser beam divergence will be lower in the vertical direction than in the horizontal.

Diffraction-limited Imaging

The Airy disk around each of the stars from the 2.56 m telescope aperture can be seen in this *lucky image* of the binary star zeta Boötis.

The ability of an imaging system to resolve detail is ultimately limited by diffraction. This is because a plane wave incident on a circular lens or mirror is diffracted as described above. The light is not focused to a point but forms an Airy disk having a central spot in the focal plane with radius to first null of

$$d = 1.22\lambda N,$$

where λ is the wavelength of the light and N is the f-number (focal length divided by diameter) of the imaging optics. In object space, the corresponding angular resolution is

$$\sin\theta = 1.22\frac{\lambda}{D},$$

where D is the diameter of the entrance pupil of the imaging lens (e.g., of a telescope's main mirror).

Two point sources will each produce an Airy pattern. As the point sources move closer together, the patterns will start to overlap, and ultimately they will merge to form a single pattern, in which case the two point sources cannot be resolved in the image. The Rayleigh criterion specifies that two point sources can be considered to be resolvable if the separation of the two images is at least the radius of the Airy disk, i.e. if the first minimum of one coincides with the maximum of the other.

Thus, the larger the aperture of the lens, and the smaller the wavelength, the finer the resolution of an imaging system. This is why telescopes have very large lenses or mirrors, and why optical microscopes are limited in the detail which they can see.

Speckle Patterns

The speckle pattern which is seen when using a laser pointer is another diffraction phenomenon. It is a result of the superposition of many waves with different phases, which are produced when a laser beam illuminates a rough surface. They add together to give a resultant wave whose amplitude, and therefore intensity, varies randomly.

Babinet's Principle

Babinet's Principle is a useful theorem stating that the diffraction pattern from an opaque body is identical to that from a hole of the same size and shape, but with differing intensities. This means that the interference conditions of a single obstruction would be the same as that of a single slit.

Patterns

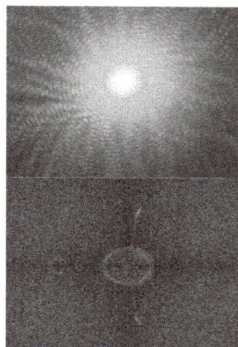

The upper half of this image shows a diffraction pattern of He-Ne laser beam on an elliptic aperture. The lower half is its 2D Fourier transform approximately reconstructing the shape of the aperture.

Several qualitative observations can be made of diffraction in general:

- The angular spacing of the features in the diffraction pattern is inversely proportional to the dimensions of the object causing the diffraction. In other words: The smaller the diffracting object, the 'wider' the resulting diffraction pattern, and vice versa. (More precisely, this is true of the sines of the angles.)

- The diffraction angles are invariant under scaling; that is, they depend only on the ratio of the wavelength to the size of the diffracting object.

- When the diffracting object has a periodic structure, for example in a diffraction grating, the features generally become sharper. The third figure, for example, shows a comparison of a double-slit pattern with a pattern formed by five slits, both sets of slits having the same spacing, between the center of one slit and the next.

Particle Diffraction

Quantum theory tells us that every particle exhibits wave properties. In particular, massive particles can interfere and therefore diffract. Diffraction of electrons and neutrons stood as one of the powerful arguments in favor of quantum mechanics. The wavelength associated with a particle is the de Broglie wavelength

$$\lambda = \frac{h}{p}$$

where h is Planck's constant and p is the momentum of the particle (mass × velocity for slow-moving particles).

For most macroscopic objects, this wavelength is so short that it is not meaningful to assign a wavelength to them. A sodium atom traveling at about 30,000 m/s would have a De Broglie wavelength of about 50 pico meters.

Because the wavelength for even the smallest of macroscopic objects is extremely small, diffraction of matter waves is only visible for small particles, like electrons, neutrons, atoms and small molecules. The short wavelength of these matter waves makes them ideally suited to study the atomic crystal structure of solids and large molecules like proteins.

Relatively larger molecules like buckyballs were also shown to diffract.

Bragg Diffraction

Diffraction from a three-dimensional periodic structure such as atoms in a crystal is called Bragg diffraction. It is similar to what occurs when waves are scattered from a diffraction grating. Bragg diffraction is a consequence of interference between waves reflecting from different crystal planes. The condition of constructive interference is given by *Bragg's law*:

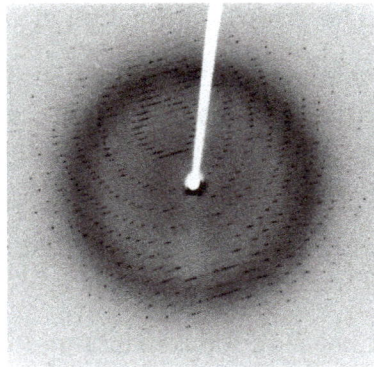

Following Bragg's law, each dot (or *reflection*) in this diffraction pattern forms from the constructive interference of X-rays passing through a crystal. The data can be used to determine the crystal's atomic structure.

$$m\lambda = 2d \sin\theta$$

where

λ is the wavelength,

d is the distance between crystal planes,

θ is the angle of the diffracted wave.

and m is an integer known as the *order* of the diffracted beam.

Bragg diffraction may be carried out using either light of very short wavelength like X-rays or matter waves like neutrons (and electrons) whose wavelength is on the order of (or much smaller than) the atomic spacing. The pattern produced gives information of the separations of crystallographic planes d, allowing one to deduce the crystal structure. Diffraction contrast, in electron microscopes and x-topography devices in particular, is also a powerful tool for examining individual defects and local strain fields in crystals.

Coherence

The description of diffraction relies on the interference of waves emanating from the same source taking different paths to the same point on a screen. In this description, the difference in phase between waves that took different paths is only dependent on the effective path length. This does not take into account the fact that waves that arrive at the screen at the same time were emitted by the source at different times. The initial phase with which the source emits waves can change over time in an unpredictable way. This means that waves emitted by the source at times that are too far apart can no longer form a constant interference pattern since the relation between their phases is no longer time independent.

The length over which the phase in a beam of light is correlated, is called the coherence length. In order for interference to occur, the path length difference must be smaller than the coherence length. This is sometimes referred to as spectral coherence, as it is related to the presence of different frequency components in the wave. In the case of light emitted by an atomic transition, the coherence length is related to the lifetime of the excited state from which the atom made its transition.

If waves are emitted from an extended source, this can lead to incoherence in the transversal direction. When looking at a cross section of a beam of light, the length over which the phase is correlated is called the transverse coherence length. In the case of Young's double slit experiment, this would mean that if the transverse coherence length is smaller than the spacing between the two slits, the resulting pattern on a screen would look like two single slit diffraction patterns.

In the case of particles like electrons, neutrons and atoms, the coherence length is related to the spatial extent of the wave function that describes the particle.

Reflection (Physics)

Reflection is the change in direction of a wavefront at an interface between two different media so that the wavefront returns into the medium from which it originated. Common examples include the reflection of light, sound and water waves. The *law of reflection* says that for specular reflection the angle at which the wave is incident on the surface equals the angle at which it is reflected. Mirrors exhibit specular reflection.

The reflection of Mount Hood in Mirror Lake.

In acoustics, reflection causes echoes and is used in sonar. In geology, it is important in the study of seismic waves. Reflection is observed with surface waves in bodies of water. Reflection is observed with many types of electromagnetic wave, besides visible light. Reflection of VHF and higher frequencies is important for radio transmission and for radar. Even hard X-rays and gamma rays can be reflected at shallow angles with special "grazing" mirrors.

Reflection of Light

Reflection of light is either *specular* (mirror-like) or *diffuse* (retaining the energy, but losing the image) depending on the nature of the interface. In specular reflection the phase of the reflected waves depends on the choice of the origin of coordinates, but the relative phase between s and p (TE and TM) polarizations is fixed by the properties of the media and of the interface between them.

A mirror provides the most common model for specular light reflection, and typically consists of a glass sheet with a metallic coating where the reflection actually occurs. Reflection is enhanced in metals by suppression of wave propagation beyond their skin depths. Reflection also occurs at the surface of transparent media, such as water or glass.

In the diagram at left, a light ray PO strikes a vertical mirror at point O, and the reflected ray is OQ. By projecting an imaginary line through point O perpendicular to the mirror, known as the *normal*, we can measure the *angle of incidence, θ_i* and the *angle of reflection, θ_r.* The *law of reflection* states that $\theta_i = \theta_r$, *or in other words, the angle of incidence equals the angle of reflection.*

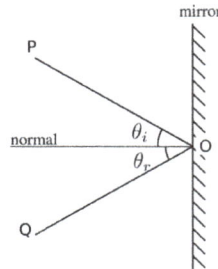

Diagram of specular reflection

In fact, reflection of light may occur whenever light travels from a medium of a given refractive index into a medium with a different refractive index. In the most general case, a certain fraction of the light is reflected from the interface, and the remainder is refracted. Solving Maxwell's equations for a light ray striking a boundary allows the derivation of the Fresnel equations, which can be used to predict how much of the light is reflected, and how much is refracted in a given situation. This is analogous to the way impedance mismatch in an electric circuit causes reflection of signals. Total internal reflection of light from a denser medium occurs if the angle of incidence is above the critical angle.

Total internal reflection is used as a means of focusing waves that cannot effectively be reflected by common means. X-ray telescopes are constructed by creating a converging "tunnel" for the waves. As the waves interact at low angle with the surface of this tunnel they are reflected toward the focus point (or toward another interaction with the tunnel surface, eventually being directed to the detector at the focus). A conventional reflector would be useless as the X-rays would simply pass through the intended reflector.

When light reflects off a material denser (with higher refractive index) than the external medium, it undergoes a polarity inversion. In contrast, a less dense, lower refractive index material will reflect light in phase. This is an important principle in the field of thin-film optics.

Specular reflection forms images. Reflection from a flat surface forms a mirror image, which appears to be reversed from left to right because we compare the image we see to what we would see if we were rotated into the position of the image. Specular reflection at a curved surface forms an image which may be magnified or demagnified; curved mirrors have optical power. Such mirrors may have surfaces that are spherical or parabolic.

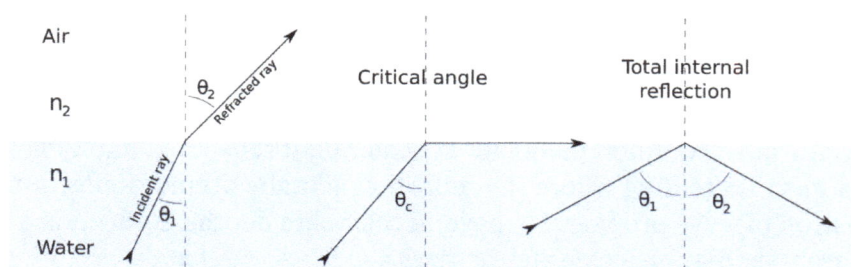

Refraction of light at the interface between two media.

Laws of Reflection

If the reflecting surface is very smooth, the reflection of light that occurs is called specular or regular reflection. The laws of reflection are as follows:

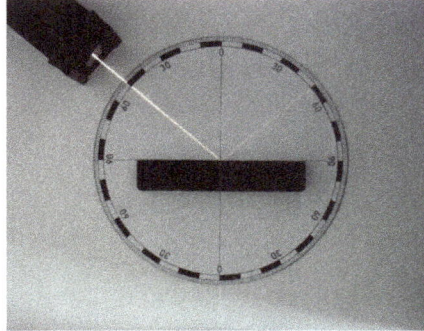

An example of the law of reflection

1. The incident ray, the reflected ray and the normal to the reflection surface at the point of the incidence lie in the same plane.

2. The angle which the incident ray makes with the normal is equal to the angle which the reflected ray makes to the same normal.

3. The reflected ray and the incident ray are on the opposite sides of the normal.

These three laws can all be derived from the Fresnel equations.

Mechanism

In classical electrodynamics, light is considered as an electromagnetic wave, which is described by Maxwell's equations. Light waves incident on a material induce small oscillations of polarisation in the individual atoms (or oscillation of electrons, in metals), causing each particle to radiate a small secondary wave in all directions, like a dipole antenna. All these waves add up to give specular reflection and refraction, according to the Huygens–Fresnel principle.

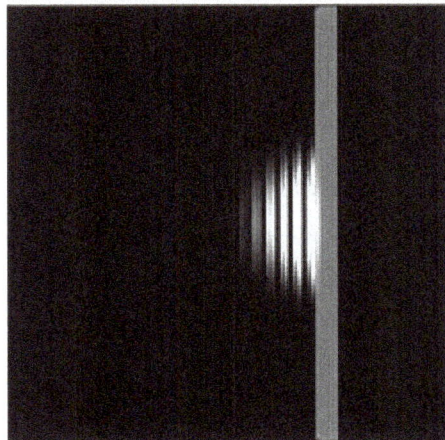

2D simulation: reflection of a quantum particle. White blur represents the probability distribution of finding a particle in a given place if measured.

In the case of dielectrics such as glass, the electric field of the light acts on the electrons in the material, and the moving electrons generate fields and become new radiators. The refracted light in the glass is the combination of the forward radiation of the electrons and the incident light. The reflected light is the combination of the backward radiation of all of the electrons.

In metals, electrons with no binding energy are called free electrons. When these electrons oscillate with the incident light, the phase difference between their radiation field and the incident field is π (180°), so the forward radiation cancels the incident light, and backward radiation is just the reflected light.

Light–matter interaction in terms of photons is a topic of quantum electrodynamics, and is described in detail by Richard Feynman in his popular book *QED: The Strange Theory of Light and Matter*.

Diffuse Reflection

When light strikes the surface of a (non-metallic) material it bounces off in all directions due to multiple reflections by the microscopic irregularities *inside* the material (e.g. the grain boundaries of a polycrystalline material, or the cell or fiber boundaries of an organic material) and by its surface, if it is rough. Thus, an 'image' is not formed. This is called *diffuse reflection*. The exact form of the reflection depends on the structure of the material. One common model for diffuse reflection is Lambertian reflectance, in which the light is reflected with equal luminance (in photometry) or radiance (in radiometry) in all directions, as defined by Lambert's cosine law.

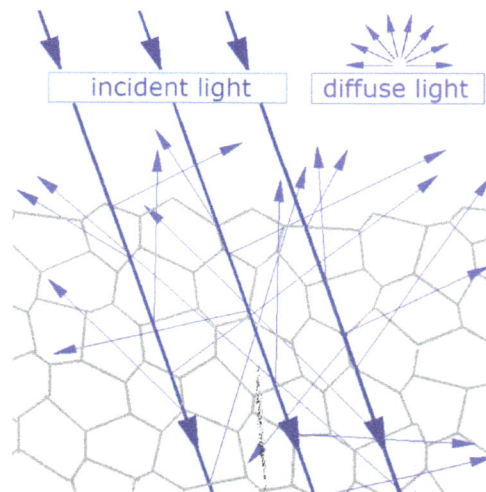

General scattering mechanism which gives diffuse reflection by a solid surface

The light sent to our eyes by most of the objects we see is due to diffuse reflection from their surface, so that this is our primary mechanism of physical observation.

Retroreflection

Some surfaces exhibit *retroreflection*. The structure of these surfaces is such that light is returned in the direction from which it came.

When flying over clouds illuminated by sunlight the region seen around the aircraft's shadow will appear brighter, and a similar effect may be seen from dew on grass. This partial retro-reflection is created by the refractive properties of the curved droplet's surface and reflective properties at the backside of the droplet.

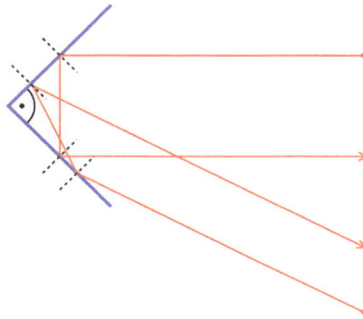

Working principle of a corner reflector

Some animals' retinas act as retroreflectors, as this effectively improves the animals' night vision. Since the lenses of their eyes modify reciprocally the paths of the incoming and outgoing light the effect is that the eyes act as a strong retroreflector, sometimes seen at night when walking in wild-lands with a flashlight.

A simple retroreflector can be made by placing three ordinary mirrors mutually perpendicular to one another (a corner reflector). The image produced is the inverse of one produced by a single mirror. A surface can be made partially retroreflective by depositing a layer of tiny refractive spheres on it or by creating small pyramid like structures. In both cases internal reflection causes the light to be reflected back to where it originated. This is used to make traffic signs and automobile license plates reflect light mostly back in the direction from which it came. In this application perfect retroreflection is not desired, since the light would then be directed back into the head-lights of an oncoming car rather than to the driver's eyes.

Multiple Reflections

When light reflects off a mirror, one image appears. Two mirrors placed exactly face to face give the appearance of an infinite number of images along a straight line. The multiple images seen between two mirrors that sit at an angle to each other lie over a circle. The center of that circle is located at the imaginary intersection of the mirrors. A square of four mirrors placed face to face give the appearance of an infinite number of images arranged in a plane. The multiple images seen between four mirrors assembling a pyramid, in which each pair of mirrors sits an angle to each other, lie over a sphere. If the base of the pyramid is rectangle shaped, the images spread over a section of a torus.

Complex Conjugate Reflection

In this process (which is also known as phase conjugation), light bounces exactly back in the di-rection from which it came due to a nonlinear optical process. Not only the direction of the light is reversed, but the actual wavefronts are reversed as well. A conjugate reflector can be used to remove aberrations from a beam by reflecting it and then passing the reflection through the aber-rating optics a second time.

Other Types of Reflection

Neutron Reflection

Materials that reflect neutrons, for example beryllium, are used in nuclear reactors and nuclear weapons. In the physical and biological sciences, the reflection of neutrons off of atoms within a material is commonly used to determine the material's internal structure.

Sound Reflection

When a longitudinal sound wave strikes a flat surface, sound is reflected in a coherent manner provided that the dimension of the reflective surface is large compared to the wavelength of the sound. Note that audible sound has a very wide frequency range (from 20 to about 17000 Hz), and thus a very wide range of wavelengths (from about 20 mm to 17 m). As a result, the overall nature of the reflection varies according to the texture and structure of the surface. For example, porous materials will absorb some energy, and rough materials (where rough is relative to the wavelength) tend to reflect in many directions—to scatter the energy, rather than to reflect it coherently. This leads into the field of architectural acoustics, because the nature of these reflections is critical to the auditory feel of a space. In the theory of exterior noise mitigation, reflective surface size mildly detracts from the concept of a noise barrier by reflecting some of the sound into the opposite direction.

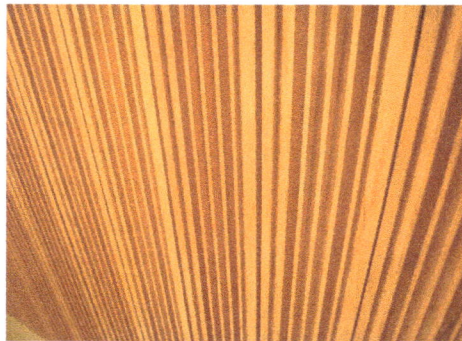

Sound diffusion panel for high frequencies

Seismic Reflection

Seismic waves produced by earthquakes or other sources (such as explosions) may be reflected by layers within the Earth. Study of the deep reflections of waves generated by earthquakes has allowed seismologists to determine the layered structure of the Earth. Shallower reflections are used in reflection seismology to study the Earth's crust generally, and in particular to prospect for petroleum and natural gas deposits.

References

- Bird, Byron; Stewart, Warren; Lightfoot, Edward (2007). Transport Phenomena. New York: Wiley, Second Edition. p. 912. ISBN 0-471-41077-2.

- Milsom, J. (2003). Field Geophysics. The geological field guide series. 25. John Wiley and Sons. p. 232. ISBN 978-0-470-84347-5. Retrieved 2010-02-25.

- Justin L Rubinstein, DR Shelly & WL Ellsworth (2009). "Non-volcanic tremor: A window into the roots of fault zones". In S. Cloetingh, Jorg Negendank. New Frontiers in Integrated Solid Earth Sciences. Springer. p. 287 ff. ISBN 90-481-2736-X. The analysis of seismic waves provides a direct high-resolution means for studying the internal structure of the Earth...

- CMR Fowler (2005). "§4.1 Waves through the Earth". The solid earth: an introduction to global geophysics (2nd ed.). Cambridge University Press. p. 100. ISBN 0-521-58409-4. Seismology is the study of the passage of elastic waves through the Earth. It is arguably the most powerful method available for studying the structure of the interior of the Earth, especially the crust and mantle.

- Halliday, David; Resnick, Robert; Walker, Jerl (2005), Fundamental of Physics (7th ed.), USA: John Wiley and Sons, Inc., ISBN 0-471-23231-9

- .Ayahiko Ichimiya; Philip I. Cohen (13 December 2004). Reflection High-Energy Electron Diffraction. Cambridge University Press. ISBN 978-0-521-45373-8.

- Lekner, John (1987). Theory of Reflection, of Electromagnetic and Particle Waves. Springer. ISBN 9789024734184.

- Arumugam, Nadia. "Food Explainer: Why Is Some Deli Meat Iridescent?". Slate. The Slate Group. Retrieved 9 September 2013.

- S. P. Näsholm and S. Holm, "On a Fractional Zener Elastic Wave Equation," Fract. Calc. Appl. Anal. Vol. 16, No 1 (2013), pp. 26-50, DOI: 10.2478/s13540-013--0003-1 Link to e-print

- Richard Feynman, Lectures in Physics, Volume 1, Chapter 47: Sound. The wave equation, Caltech 1963, 2006, 2013

4

Essential Aspects of Acoustics

The major aspects of acoustics are discussed in this chapter. Vibration, sound, ultrasound and infrasound are some of the significant topics related to acoustics. The following chapter unfolds its crucial aspects in a critical yet systematic manner.

Vibration

Vibration is a mechanical phenomenon whereby oscillations occur about an equilibrium point. The word comes from Latin *vibrationem* ("shaking, brandishing"). The oscillations may be periodic, such as the motion of a pendulum—or random, such as the movement of a tire on a gravel road.

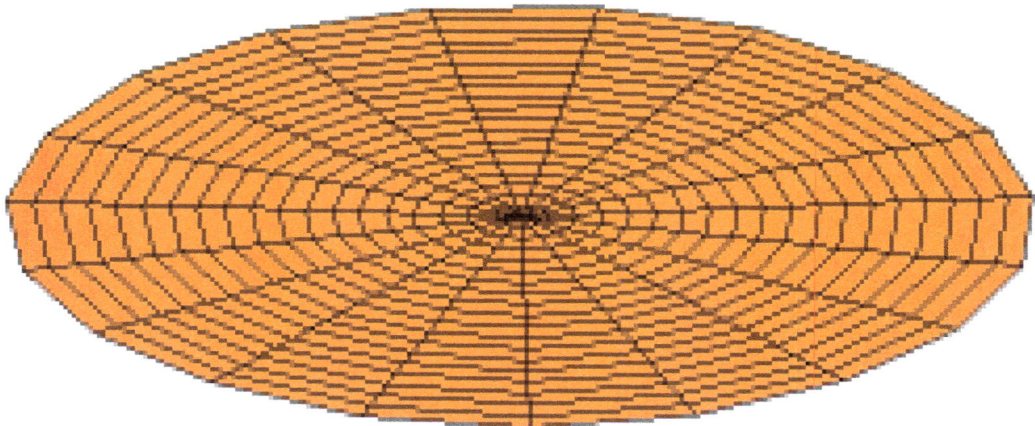

One of the possible modes of vibration of a circular drum.

Vibration can be desirable: for example, the motion of a tuning fork, the reed in a woodwind instrument or harmonica, a mobile phone, or the cone of a loudspeaker.

In many cases, however, vibration is undesirable, wasting energy and creating unwanted sound. For example, the vibrational motions of engines, electric motors, or any mechanical device in operation are typically unwanted. Such vibrations could be caused by imbalances in the rotating parts, uneven friction, or the meshing of gear teeth. Careful designs usually minimize unwanted vibrations.

The studies of sound and vibration are closely related. Sound, or pressure waves, are generated by vibrating structures (e.g. vocal cords); these pressure waves can also induce the vibration of structures (e.g. ear drum). Hence, attempts to reduce noise are often related to issues of vibration.

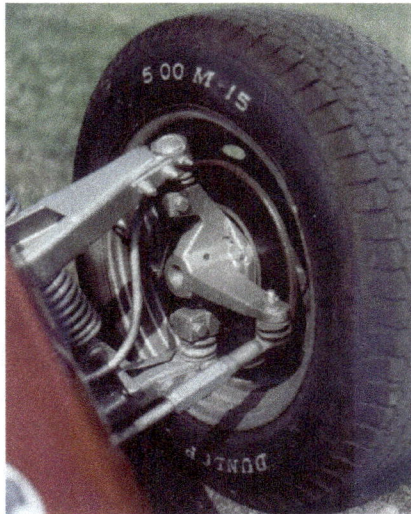

Car Suspension: designing vibration control is undertaken as part of acoustic, automotive or mechanical engineering.

Types of Vibration

Free vibration occurs when a mechanical system is set in motion with an initial input and allowed to vibrate freely. Examples of this type of vibration are pulling a child back on a swing and letting go, or hitting a tuning fork and letting it ring. The mechanical system vibrates at one or more of its natural frequencies and damps down to motionlessness.

Forced vibration is when a time-varying disturbance (load, displacement or velocity) is applied to a mechanical system. The disturbance can be a periodic and steady-state input, a transient input, or a random input. The periodic input can be a harmonic or a non-harmonic disturbance. Examples of these types of vibration include a washing machine shaking due to an imbalance, transportation vibration caused by an engine or uneven road, or the vibration of a building during an earthquake. For linear systems, the frequency of the steady-state vibration response resulting from the application of a periodic, harmonic input is equal to the frequency of the applied force or motion, with the response magnitude being dependent on the actual mechanical system.

Vibration Testing

Vibration testing is accomplished by introducing a forcing function into a structure, usually with some type of shaker. Alternately, a DUT (device under test) is attached to the "table" of a shaker. Vibration testing is performed to examine the response of a device under test (DUT) to a defined vibration environment. The measured response may be fatigue life, resonant frequencies or squeak and rattle sound output (NVH). Squeak and rattle testing is performed with a special type of *quiet shaker* that produces very low sound levels while under operation.

For relatively low frequency forcing, servohydraulic (electrohydraulic) shakers are used. For higher frequencies, electrodynamic shakers are used. Generally, one or more "input" or "control" points located on the DUT-side of a fixture is kept at a specified acceleration. Other "response" points experience maximum vibration level (resonance) or minimum vibration level (anti-resonance). It is often desirable to achieve anti-resonance to keep a system from becoming too noisy, or to reduce strain on certain parts due to vibration modes caused by specific vibration frequencies .

The most common types of vibration testing services conducted by vibration test labs are Sinusoidal and Random. Sine (one-frequency-at-a-time) tests are performed to survey the structural response of the device under test (DUT). A random (all frequencies at once) test is generally considered to more closely replicate a real world environment, such as road inputs to a moving automobile.

Most vibration testing is conducted in a 'single DUT axis' at a time, even though most real-world vibration occurs in various axes simultaneously. MIL-STD-810G, released in late 2008, Test Method 527, calls for multiple exciter testing. The *vibration test fixture* used to attach the DUT to the shaker table must be designed for the frequency range of the vibration test spectrum. Generally for smaller fixtures and lower frequency ranges, the designer targets a fixture design that is free of resonances in the test frequency range. This becomes more difficult as the DUT gets larger and as the test frequency increases. In these cases multi-point control strategies can mitigate some of the resonances that may be present in the future. Devices specifically designed to trace or record vibrations are called vibroscopes.

Vibration Analysis

Vibration Analysis (VA), applied in an industrial or maintenance environment aims to reduce maintenance costs and equipment downtime by detecting equipment faults. VA is a key component of a Condition Monitoring (CM) program, and is often referred to as Predictive Maintenance (PdM). Most commonly VA is used to detect faults in rotating equipment (Fans, Motors, Pumps, and Gearboxes etc.) such as Unbalance, Misalignment, rolling element bearing faults and resonance conditions.

VA can use the units of Displacement, Velocity and Acceleration displayed as a Time Waveform (TWF), but most commonly the spectrum is used, derived from a Fast Fourier Transform of the TWF. The vibration spectrum provides important frequency information that can pinpoint the faulty component.

The fundamentals of vibration analysis can be understood by studying the simple mass–spring–damper model. Indeed, even a complex structure such as an automobile body can be modeled as a "summation" of simple mass–spring–damper models. The mass–spring–damper model is an example of a simple harmonic oscillator. The mathematics used to describe its behavior is identical to other simple harmonic oscillators such as the RLC circuit.

Note: This article does not include the step-by-step mathematical derivations, but focuses on major vibration analysis equations and concepts. Please refer to the references at the end of the article for detailed derivations.

Free Vibration Without Damping

To start the investigation of the mass–spring–damper assume the damping is negligible and that there is no external force applied to the mass (i.e. free vibration). The force applied to the mass by the spring is proportional to the amount the spring is stretched "x" (assuming the spring is already compressed due to the weight of the mass). The proportionality constant, k, is the stiffness of the spring and has units of force/distance (e.g. lbf/in or N/m). The negative sign indicates that the force is always opposing the motion of the mass attached to it:

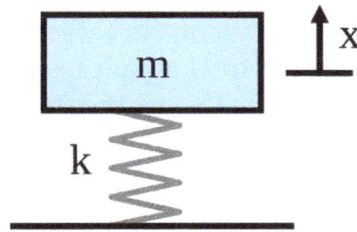

$$F_s = -kx.$$

The force generated by the mass is proportional to the acceleration of the mass as given by Newton's second law of motion :

$$\Sigma F = ma = m\ddot{x} = m\frac{d^2x}{dt^2}.$$

The sum of the forces on the mass then generates this ordinary differential equation: $m\ddot{x} + kx = 0$.

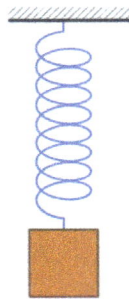

Simple harmonic motion of the mass–spring system

Assuming that the initiation of vibration begins by stretching the spring by the distance of A and releasing, the solution to the above equation that describes the motion of mass is:

$$x(t) = A\cos(2\pi f_n t).$$

This solution says that it will oscillate with simple harmonic motion that has an amplitude of A and a frequency of f_n. The number f_n is called the undamped natural frequency. For the simple mass–spring system, f_n is defined as:

$$f_n = \frac{1}{2\pi}\sqrt{\frac{k}{m}}.$$

Note: angular frequency ω ($\omega = 2\pi f$) with the units of radians per second is often used in equations because it simplifies the equations, but is normally converted to ordinary frequency (units of Hz or equivalently cycles per second) when stating the frequency of a system. If the mass and stiffness of the system is known, the formula above can determine the frequency at which the system vibrates once set in motion by an initial disturbance. Every vibrating system has one or more natural frequencies that it vibrates at once disturbed. This simple relation can be used to understand in general what happens to a more complex system once we add mass or stiffness. For example, the above formula explains why, when a car or truck is fully loaded, the suspension feels "softer" than unloaded—the mass has increased, reducing the natural frequency of the system.

What Causes The System to Vibrate: from Conservation of Energy Point of View

Vibrational motion could be understood in terms of conservation of energy. In the above example the spring has been extended by a value of x and therefore some potential energy ($\frac{1}{2}kx^2$) is stored in the spring. Once released, the spring tends to return to its un-stretched state (which is the minimum potential energy state) and in the process accelerates the mass. At the point where the spring has reached its un-stretched state all the potential energy that we supplied by stretching it has been transformed into kinetic energy ($\frac{1}{2}mv^2$). The mass then begins to decelerate because it is now compressing the spring and in the process transferring the kinetic energy back to its potential. Thus oscillation of the spring amounts to the transferring back and forth of the kinetic energy into potential energy. In this simple model the mass continues to oscillate forever at the same magnitude—but in a real system, *damping* always dissipates the energy, eventually bringing the spring to rest.

Free Vibration with Damping

When a "viscous" damper is added to the model this outputs a force that is proportional to the velocity of the mass. The damping is called viscous because it models the effects of a fluid within an object. The proportionality constant c is called the damping coefficient and has units of Force over velocity (lbf s/ in or N s/m).

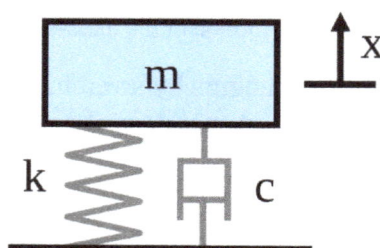

Mass Spring Damper Model

$$F_d = -cv = -c\dot{x} = -c\frac{dx}{dt}.$$

Summing the forces on the mass results in the following ordinary differential equation:

$$m\ddot{x} + c\dot{x} + kx = 0.$$

The solution to this equation depends on the amount of damping. If the damping is small enough, the system still vibrates—but eventually, over time, stops vibrating. This case is called underdamping, which is important in vibration analysis. If damping is increased just to the point where the system no longer oscillates, the system has reached the point of *critical damping*. If the damping is increased past critical damping, the system is *overdamped*. The value that the damping coefficient must reach for critical damping in the mass spring damper model is:

$$c_c = 2\sqrt{km}.$$

To characterize the amount of damping in a system a ratio called the damping ratio (also known as damping factor and % critical damping) is used. This damping ratio is just a ratio of the actual damping over the amount of damping required to reach critical damping. The formula for the damping ratio (ζ) of the mass spring damper model is:

$$\zeta = \frac{c}{2\sqrt{km}}.$$

For example, metal structures (e.g., airplane fuselages, engine crankshafts) have damping factors less than 0.05, while automotive suspensions are in the range of 0.2–0.3. The solution to the underdamped system for the mass spring damper model is the following:

$$x(t) = Xe^{-\zeta\omega_n t}\cos(\sqrt{1-\zeta^2}\,\omega_n t - \phi), \qquad \omega_n = 2\pi f_n.$$

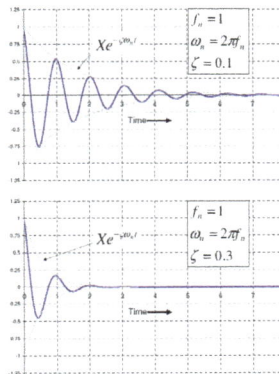

Free vibration with 0.1 and 0.3 damping ratio

The value of X, the initial magnitude, and the phase shift, are determined by the amount the spring is stretched. The formulas for these values can be found in the references.

Damped and Undamped Natural Frequencies

The major points to note from the solution are the exponential term and the cosine function. The exponential term defines how quickly the system "damps" down – the larger the damping ratio, the quicker it damps to zero. The cosine function is the oscillating portion of the solution, but the frequency of the oscillations is different from the undamped case.

The frequency in this case is called the "damped natural frequency", f_d, and is related to the undamped natural frequency by the following formula:

$$f_d = f_n \sqrt{1 - \zeta^2}.$$

The damped natural frequency is less than the undamped natural frequency, but for many practical cases the damping ratio is relatively small and hence the difference is negligible. Therefore, the damped and undamped description are often dropped when stating the natural frequency (e.g. with 0.1 damping ratio, the damped natural frequency is only 1% less than the undamped).

The plots to the side present how 0.1 and 0.3 damping ratios effect how the system "rings" down over time. What is often done in practice is to experimentally measure the free vibration after an impact (for example by a hammer) and then determine the natural frequency of the system by measuring the rate of oscillation, as well as the damping ratio by measuring the rate of decay. The natural frequency and damping ratio are not only important in free vibration, but also characterize how a system behaves under forced vibration.

Forced Vibration with Damping

The behavior of the spring mass damper model varies with the addition of a harmonic force. A force of this type could, for example, be generated by a rotating imbalance.

$$F = F_0 \sin(2\pi ft).$$

Summing the forces on the mass results in the following ordinary differential equation:

$$m\ddot{x} + c\dot{x} + kx = F_0 \sin(2\pi ft).$$

The steady state solution of this problem can be written as:

$$x(t) = X \sin(2\pi ft + \phi).$$

The result states that the mass will oscillate at the same frequency, f, of the applied force, but with a phase shift

The amplitude of the vibration "X" is defined by the following formula.

$$X = \frac{F_0}{k} \frac{1}{\sqrt{(1 - r^2)^2 + (2\zeta r)^2}}.$$

Where "r" is defined as the ratio of the harmonic force frequency over the undamped natural frequency of the mass–spring–damper model.

$$r = \frac{f}{f_n}.$$

The phase shift, ϕ, is defined by the following formula.

$$\phi = \arctan\left(\frac{2\zeta r}{1 - r^2}\right).$$

If resonance occurs in a mechanical system it can be very harmful – leading to eventual failure of the system. Consequently, one of the major reasons for vibration analysis is to predict when this type of resonance may occur and then to determine what steps to take to prevent it from occurring. As the amplitude plot shows, adding damping can significantly reduce the magnitude of the vibration. Also, the magnitude can be reduced if the natural frequency can be shifted away from the forcing frequency by changing the stiffness or mass of the system. If the system cannot be changed, perhaps the forcing frequency can be shifted (for example, changing the speed of the machine generating the force).

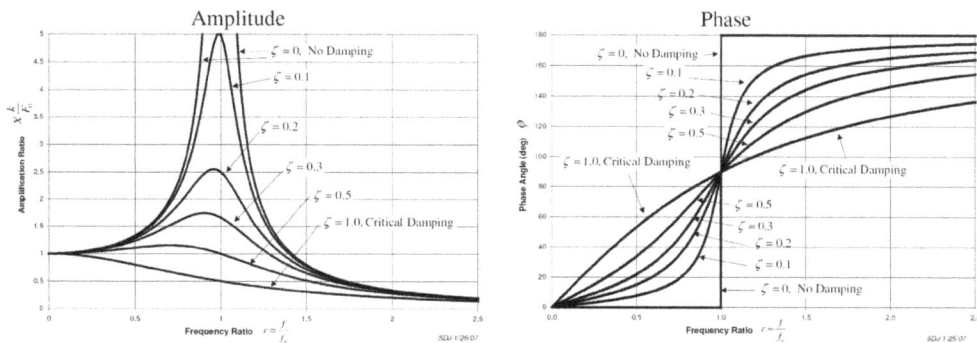

The plot of these functions, called "the frequency response of the system", presents one of the most important features in forced vibration. In a lightly damped system when the forcing frequency nears the natural frequency ($r \approx 1$) the amplitude of the vibration can get extremely high. This phenomenon is called resonance (subsequently the natural frequency of a system is often referred to as the resonant frequency). In rotor bearing systems any rotational speed that excites a resonant frequency is referred to as a critical speed.

The following are some other points in regards to the forced vibration shown in the frequency response plots.

- At a given frequency ratio, the amplitude of the vibration, X, is directly proportional to the amplitude of the force F_0 (e.g. if you double the force, the vibration doubles)

- With little or no damping, the vibration is in phase with the forcing frequency when the frequency ratio $r < 1$ and 180 degrees out of phase when the frequency ratio $r > 1$

- When $r \ll 1$ the amplitude is just the deflection of the spring under the static force F_0. This deflection is called the static deflection δ_{st}. Hence, when $r \ll 1$ the effects of the damper and the mass are minimal.

- When $r \gg 1$ the amplitude of the vibration is actually less than the static deflection δ_{st}. In this region the force generated by the mass ($F = ma$) is dominating because the acceleration seen by the mass increases with the frequency. Since the deflection seen in the spring, X, is reduced in this region, the force transmitted by the spring ($F = kx$) to the base is reduced. Therefore, the mass–spring–damper system is isolating the harmonic force from the mounting base – referred to as vibration isolation. Interestingly, more damping actually reduces the effects of vibration isolation when $r \gg 1$ because the damping force ($F = cv$) is also transmitted to the base.

- whatever the damping is, the vibration is 90 degrees out of phase with the forcing frequency when the frequency ratio r =1, which is very helpful when it comes to determining the natural frequency of the system.

- whatever the damping is, when r ≫ 1, the vibration is 180 degrees out of phase with the forcing frequency

- whatever the damping is, when r ≪ 1, the vibration is in phase with the forcing frequency

What Causes Resonance?

Resonance is simple to understand if the spring and mass are viewed as energy storage elements – with the mass storing kinetic energy and the spring storing potential energy. As discussed earlier, when the mass and spring have no external force acting on them they transfer energy back and forth at a rate equal to the natural frequency. In other words, to efficiently pump energy into both mass and spring requires that the energy source feed the energy in at a rate equal to the natural frequency. Applying a force to the mass and spring is similar to pushing a child on swing, a push is needed at the correct moment to make the swing get higher and higher. As in the case of the swing, the force applied need not be high to get large motions, but must just add energy to the system.

The damper, instead of storing energy, dissipates energy. Since the damping force is proportional to the velocity, the more the motion, the more the damper dissipates the energy. Therefore, there is a point when the energy dissipated by the damper equals the energy added by the force. At this point, the system has reached its maximum amplitude and will continue to vibrate at this level as long as the force applied stays the same. If no damping exists, there is nothing to dissipate the energy and, theoretically, the motion will continue to grow into infinity.

Applying "Complex" Forces to The Mass–spring–damper Model

In a previous section only a simple harmonic force was applied to the model, but this can be extended considerably using two powerful mathematical tools. The first is the Fourier transform that takes a signal as a function of time (time domain) and breaks it down into its harmonic components as a function of frequency (frequency domain). For example, by applying a force to the mass–spring–damper model that repeats the following cycle – a force equal to 1 newton for 0.5 second and then no force for 0.5 second. This type of force has the shape of a 1 Hz square wave.

The Fourier transform of the square wave generates a frequency spectrum that presents the magnitude of the harmonics that make up the square wave (the phase is also generated, but is typically of less concern and therefore is often not plotted). The Fourier transform can also be used to analyze non-periodic functions such as transients (e.g. impulses) and random functions. The Fourier transform is almost always computed using the Fast Fourier Transform (FFT) computer algorithm in combination with a window function.

In the case of our square wave force, the first component is actually a constant force of 0.5 newton and is represented by a value at "0" Hz in the frequency spectrum. The next component is a 1 Hz sine wave with an amplitude of 0.64. This is shown by the line at 1 Hz. The remaining components are at odd frequencies and it takes an infinite amount of sine waves to generate the perfect square wave. Hence, the Fourier transform allows you to interpret the force as a sum of sinusoidal forces being applied instead of a more "complex" force (e.g. a square wave).

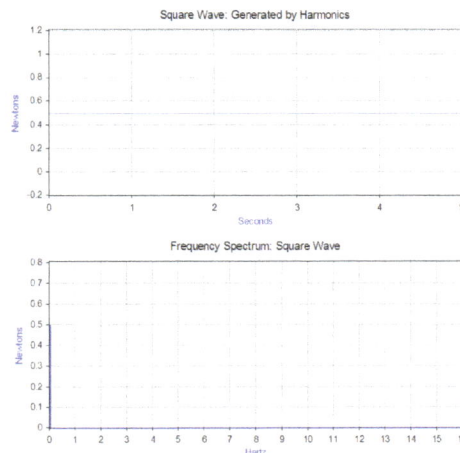

How a 1 Hz square wave can be represented as a summation of sine waves(harmonics) and the corresponding frequency spectrum. Click and go to full resolution for an animation

In the previous section, the vibration solution was given for a single harmonic force, but the Fourier transform in general gives multiple harmonic forces. The second mathematical tool, "the principle of superposition", allows the summation of the solutions from multiple forces if the system is linear. In the case of the spring–mass–damper model, the system is linear if the spring force is proportional to the displacement and the damping is proportional to the velocity over the range of motion of interest. Hence, the solution to the problem with a square wave is summing the predicted vibration from each one of the harmonic forces found in the frequency spectrum of the square wave.

Frequency Response Model

The solution of a vibration problem can be viewed as an input/output relation – where the force is the input and the output is the vibration. Representing the force and vibration in the frequency domain (magnitude and phase) allows the following relation:

$$X(i\omega) = H(i\omega) \cdot F(i\omega) \ \ or \ \ H(i\omega) = \frac{X(i\omega)}{F(i\omega)}.$$

$H(i\omega)$ is called the frequency response function (also referred to as the transfer function, but not technically as accurate) and has both a magnitude and phase component (if represented as a complex number, a real and imaginary component). The magnitude of the frequency response function (FRF) was presented earlier for the mass–spring–damper system.

$$|H(i\omega)| = \left|\frac{X(i\omega)}{F(i\omega)}\right| = \frac{1}{k} \frac{1}{\sqrt{(1-r^2)^2 + (2\zeta r)^2}}, \ \ where \ \ r = \frac{f}{f_n} = \frac{\omega}{\omega_n}.$$

The phase of the FRF was also presented earlier as:

$$\angle H(i\omega) = -\arctan\left(\frac{2\zeta r}{1-r^2}\right).$$

For example, calculating the FRF for a mass–spring–damper system with a mass of 1 kg, spring stiffness of 1.93 N/mm and a damping ratio of 0.1. The values of the spring and mass give a natural frequency of 7 Hz for this specific system. Applying the 1 Hz square wave from earlier allows the calculation of the predicted vibration of the mass. The figure illustrates the resulting vibration. It happens in this example that the fourth harmonic of the square wave falls at 7 Hz. The frequency response of the mass–spring–damper therefore outputs a high 7 Hz vibration even though the input force had a relatively low 7 Hz harmonic. This example highlights that the resulting vibration is dependent on both the forcing function and the system that the force is applied to.

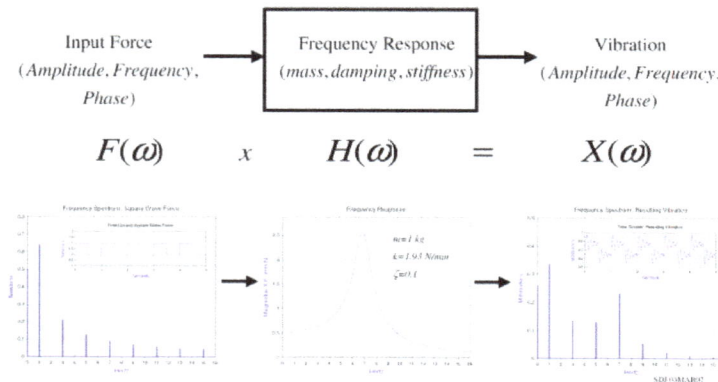

Frequency response model

The figure also shows the time domain representation of the resulting vibration. This is done by performing an inverse Fourier Transform that converts frequency domain data to time domain. In practice, this is rarely done because the frequency spectrum provides all the necessary information.

The frequency response function (FRF) does not necessarily have to be calculated from the knowledge of the mass, damping, and stiffness of the system—but can be measured experimentally. For example, if a known force over a range of frequencies is applied, and if the associated vibrations are measured, the frequency response function can be calculated, thereby characterizing the system. This technique is used in the field of experimental modal analysis to determine the vibration characteristics of a structure.

Multiple Degrees of Freedom Systems and Mode Shapes

The simple mass–spring damper model is the foundation of vibration analysis, but what about more complex systems? The mass–spring–damper model described above is called a single degree of freedom (SDOF) model since the mass is assumed to only move up and down. In more complex systems, the system must be discretized into more masses that move in more than one direction, adding degrees of freedom. The major concepts of multiple degrees of freedom (MDOF) can be understood by looking at just a 2 degree of freedom model as shown in the figure.

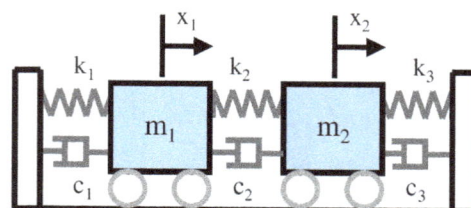

2 degree of freedom model

The equations of motion of the 2DOF system are found to be:

$$m_1 \ddot{x}_1 + (c_1 + c_2) \dot{x}_1 - c_2 \dot{x}_2 + (k_1 + k_2) x_1 - k_2 x_2 = f_1,$$

$$m_2 \ddot{x}_2 - c_2 \dot{x}_1 + (c_2 + c_3) \dot{x}_2 - k_2 x_1 + (k_2 + k_3) x_2 = f_2.$$

This can be rewritten in matrix format:

$$\begin{bmatrix} m_1 & 0 \\ 0 & m_2 \end{bmatrix} \begin{Bmatrix} \ddot{x}_1 \\ \ddot{x}_2 \end{Bmatrix} + \begin{bmatrix} c_1 + c_2 & -c_2 \\ -c_2 & c_2 + c_3 \end{bmatrix} \begin{Bmatrix} \dot{x}_1 \\ \dot{x}_2 \end{Bmatrix} + \begin{bmatrix} k_1 + k_2 & -k_2 \\ -k_2 & k_2 + k_3 \end{bmatrix} \begin{Bmatrix} x_1 \\ x_2 \end{Bmatrix} = \begin{Bmatrix} f_1 \\ f_2 \end{Bmatrix}.$$

A more compact form of this matrix equation can be written as:

$$[M]\{\ddot{x}\} + [C]\{\dot{x}\} + [K]\{x\} = \{f\}$$

where $[M]$, $[C]$, and $[K]$ are symmetric matrices referred respectively as the mass, damping, and stiffness matrices. The matrices are NxN square matrices where N is the number of degrees of freedom of the system.

In the following analysis involves the case where there is no damping and no applied forces (i.e. free vibration). The solution of a viscously damped system is somewhat more complicated.

$$[M]\{\ddot{x}\} + [K]\{x\} = 0.$$

This differential equation can be solved by assuming the following type of solution:

$$\{x\} = \{X\} e^{i\omega t}.$$

Note: Using the exponential solution of $\{X\} e^{i\omega t}$ is a mathematical trick used to solve linear differential equations. Using Euler's formula and taking only the real part of the solution it is the same cosine solution for the 1 DOF system. The exponential solution is only used because it is easier to manipulate mathematically.

The equation then becomes:

$$\left[-\omega^2 [M] + [K] \right] \{X\} e^{i\omega t} = 0.$$

Since $e^{i\omega t}$ cannot equal zero the equation reduces to the following.

$$\left[[K] - \omega^2 [M] \right] \{X\} = 0.$$

Eigenvalue Problem

This is referred to an eigenvalue problem in mathematics and can be put in the standard format by pre-multiplying the equation by $[M]^{-1}$

$$\left[[M]^{-1}[K] - \omega^2 [M]^{-1}[M] \right] \{X\} = 0$$

and if: $[M]^{-1}[K] = [A]$ and $\lambda = \omega^2$

$$\left[[A] - \lambda [I] \right] \{X\} = 0.$$

The solution to the problem results in N eigenvalues (i.e. $\omega_1^2, \omega_2^2, \cdots \omega_N^2$), where N corresponds to the number of degrees of freedom. The eigenvalues provide the natural frequencies of the system. When these eigenvalues are substituted back into the original set of equations, the values of $\{X\}$ that correspond to each eigenvalue are called the eigenvectors. These eigenvectors represent the mode shapes of the system. The solution of an eigenvalue problem can be quite cumbersome (especially for problems with many degrees of freedom), but fortunately most math analysis programs have eigenvalue routines.

The eigenvalues and eigenvectors are often written in the following matrix format and describe the modal model of the system:

$$\left[\diagdown \omega_r^2 \diagdown \right] = \begin{bmatrix} \omega_1^2 & \cdots & 0 \\ \vdots & \ddots & \vdots \\ 0 & \cdots & \omega_N^2 \end{bmatrix} \text{ and } [\Psi] = \left[\{\psi_1\}\{\psi_2\} \cdots \{\psi_N\} \right].$$

A simple example using the 2 DOF model can help illustrate the concepts. Let both masses have a mass of 1 kg and the stiffness of all three springs equal 1000 N/m. The mass and stiffness matrix for this problem are then:

$$[M] = \begin{bmatrix} 1 & 0 \\ 0 & 1 \end{bmatrix} \text{ and } [K] = \begin{bmatrix} 2000 & -1000 \\ -1000 & 2000 \end{bmatrix}.$$

$$\text{Then } [A] = \begin{bmatrix} 2000 & -1000 \\ -1000 & 2000 \end{bmatrix}.$$

The eigenvalues for this problem given by an eigenvalue routine is:

$$\left[\diagdown \omega_r^2 \diagdown \right] = \begin{bmatrix} 1000 & 0 \\ 0 & 3000 \end{bmatrix}.$$

The natural frequencies in the units of hertz are then (remembering $\omega = 2\pi f$) $f_1 = 5.033$ Hz and $f_2 = 8.717$ Hz.

The two mode shapes for the respective natural frequencies are given as:

$$[\Psi] = \left[\{\psi_1\}\{\psi_2\} \right] = \left[\begin{Bmatrix} -0.707 \\ -0.707 \end{Bmatrix}_1 \begin{Bmatrix} 0.707 \\ -0.707 \end{Bmatrix}_2 \right].$$

Since the system is a 2 DOF system, there are two modes with their respective natural frequencies and shapes. The mode shape vectors are not the absolute motion, but just describe relative motion of the degrees of freedom. In our case the first mode shape vector is saying that the masses are moving together in phase since they have the same value and sign. In the case of the second mode shape vector, each mass is moving in opposite direction at the same rate.

Illustration of A Multiple DOF Problem

When there are many degrees of freedom, one method of visualizing the mode shapes is by animating them. An example of animated mode shapes is shown in the figure below for a cantilevered I-beam. In this case, the finite element method was used to generate an approximation to the mass and stiffness matrices and solve a discrete eigenvalue problem. Note that, in this case, the finite element method provides an approximation of the 3D electrodynamics model (for which there exists an infinite number of vibration modes and frequencies). Therefore, this relatively simple model that has over 100 degrees of freedom and hence as many natural frequencies and mode shapes, provides a good approximation for the first natural frequencies and modes[†]. Generally, only the first few modes are important for practical applications.

In this table the first and second (top and bottom respectively) horizontal bending (left), torsional (middle), and vertical bending (right) vibrational modes of an I-beam are visualized. There also exist other kinds of vibrational modes in which the beam gets compressed/stretched out in the height, width and length directions respectively.

The mode shapes of a cantilevered I-beam

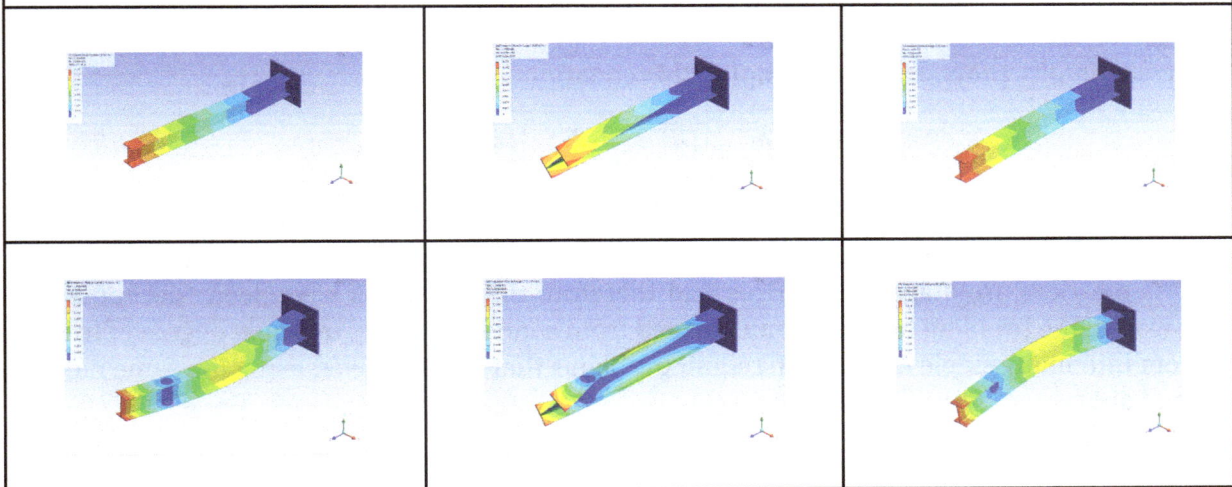

Note that when performing a numerical approximation of any mathematical model, convergence of the parameters of interest must be ascertained.

Multiple DOF Problem Converted to A Single DOF Problem

The eigenvectors have very important properties called orthogonality properties. These properties can be used to greatly simplify the solution of multi-degree of freedom models. It can be shown that the eigenvectors have the following properties:

$$[\Psi]^T [M][\Psi] = \left[\diagdown m_r \diagdown \right],$$

$$[\Psi]^T [K][\Psi] = \left[\diagdown k_r \diagdown \right].$$

$\left[\diagdown m_r \diagdown\right]$ and $\left[\diagdown k_r \diagdown\right]$ are diagonal matrices that contain the modal mass and stiffness values for each one of the modes. (Note: Since the eigenvectors (mode shapes) can be arbitrarily scaled, the orthogonality properties are often used to scale the eigenvectors so the modal mass value for each mode is equal to 1. The modal mass matrix is therefore an identity matrix)

These properties can be used to greatly simplify the solution of multi-degree of freedom models by making the following coordinate transformation.

$$\{x\} = [\Psi]\{q\}.$$

Using this coordinate transformation in the original free vibration differential equation results in the following equation.

$$[M][\Psi]\{\ddot{q}\} + [K][\Psi]\{q\} = 0.$$

Taking advantage of the orthogonality properties by premultiplying this equation by $[\varnothing]^T$

$$[\Psi]^T[M][\Psi]\{\ddot{q}\} + [\Psi]^T[K][\Psi]\{q\} = 0.$$

The orthogonality properties then simplify this equation to:

$$\left[\diagdown m_r \diagdown\right]\{\ddot{q}\} + \left[\diagdown k_r \diagdown\right]\{q\} = 0.$$

This equation is the foundation of vibration analysis for multiple degree of freedom systems. A similar type of result can be derived for damped systems. The key is that the modal mass and stiffness matrices are diagonal matrices and therefore the equations have been "decoupled". In other words, the problem has been transformed from a large unwieldy multiple degree of freedom problem into many single degree of freedom problems that can be solved using the same methods outlined above.

Solving for x is replaced by solving for q, referred to as the modal coordinates or modal participation factors.

It may be clearer to understand if $\{x\} = [\Psi]\{q\}$ is written as:

$$\{x_n\} = q_1\{\psi\}_1 + q_2\{\psi\}_2 + q_3\{\psi\}_3 + \cdots + q_N\{\psi\}_N.$$

Written in this form it can be seen that the vibration at each of the degrees of freedom is just a linear sum of the mode shapes. Furthermore, how much each mode "participates" in the final vibration is defined by q, its modal participation factor.

Sound

In physics, sound is a vibration that propagates as a typically audible mechanical wave of pressure and displacement, through a medium such as air or water. In physiology and psychology, sound is the reception of such waves and their perception by the brain.

A drum produces sound via a vibrating membrane.

Acoustics

Acoustics is the interdisciplinary science that deals with the study of mechanical waves in gases, liquids, and solids including vibration, sound, ultrasound, and infrasound. A scientist who works in the field of acoustics is an *acoustician*, while someone working in the field of acoustical engineering may be called an *acoustical engineer*. An audio engineer, on the other hand is concerned with the recording, manipulation, mixing, and reproduction of sound.

Applications of acoustics are found in almost all aspects of modern society, subdisciplines include aeroacoustics, audio signal processing, architectural acoustics, bioacoustics, electro-acoustics, environmental noise, musical acoustics, noise control, psychoacoustics, speech, ultrasound, underwater acoustics, and vibration.

Definition

Sound is defined by ANSI/ASA S1.1-2013 as "(a) Oscillation in pressure, stress, particle displacement, particle velocity, etc., propagated in a medium with internal forces (e.g., elastic or viscous), or the superposition of such propagated oscillation. (b) Auditory sensation evoked by the oscillation described in (a)."

Physics of Sound

Sound can propagate through a medium such as air, water and solids as longitudinal waves and also as a transverse wave in solids. The sound waves are generated by a sound source, such as the vibrating diaphragm of a stereo speaker. The sound source creates vibrations in the surrounding medium. As the source continues to vibrate the medium, the vibrations propagate away from the source at the speed of sound, thus forming the sound wave. At a fixed distance from the source, the pressure, velocity, and displacement of the medium vary in time. At an instant in time, the pressure, velocity, and displacement vary in space. Note that the particles of the medium do not travel with the sound wave. This is intuitively obvious for a solid, and the same is true for liquids and gases (that is, the vibrations of particles in the gas or liquid transport the vibrations, while the *average* position of the particles over time does not change). During propagation, waves can be reflected, refracted, or attenuated by the medium.

The behavior of sound propagation is generally affected by three things:

- A complex relationship between the density and pressure of the medium. This relationship, affected by temperature, determines the speed of sound within the medium.

- Motion of the medium itself. If the medium is moving, this movement may increase or decrease the absolute speed of the sound wave depending on the direction of the movement. For example, sound moving through wind will have its speed of propagation increased by the speed of the wind if the sound and wind are moving in the same direction. If the sound and wind are moving in opposite directions, the speed of the sound wave will be decreased by the speed of the wind.

- The viscosity of the medium. Medium viscosity determines the rate at which sound is attenuated. For many media, such as air or water, attenuation due to viscosity is negligible.

When sound is moving through a medium that does not have constant physical properties, it may be refracted (either dispersed or focused).

Spherical compression (longitudinal) waves

The mechanical vibrations that can be interpreted as sound are able to travel through all forms of matter: gases, liquids, solids, and plasmas. The matter that supports the sound is called the medium. Sound cannot travel through a vacuum.

Longitudinal and Transverse Waves

Sound is transmitted through gases, plasma, and liquids as longitudinal waves, also called compression waves. It requires a medium to propagate. Through solids, however, it can be transmitted as both longitudinal waves and transverse waves. Longitudinal sound waves are waves of alternating pressure deviations from the equilibrium pressure, causing local regions of compression and rarefaction, while transverse waves (in solids) are waves of alternating shear stress at right angle to the direction of propagation.

Sound waves may be "viewed" using parabolic mirrors and objects that produce sound.

The energy carried by an oscillating sound wave converts back and forth between the potential energy of the extra compression (in case of longitudinal waves) or lateral displacement strain (in case of transverse waves) of the matter, and the kinetic energy of the displacement velocity of particles of the medium.

Sound Wave Properties and Characteristics

Although there are many complexities relating to the transmission of sounds, at the point of reception (i.e. the ears), sound is readily dividable into two simple elements: pressure and time. These fundamental elements form the basis of all sound waves. They can be used to describe, in absolute terms, every sound we hear. Figure 1 shows a 'pressure over time' graph of a 20 ms recording of a clarinet tone).

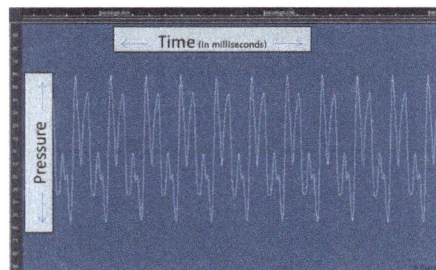

Figure 1. the two fundamental elements of sound; Pressure and Time.

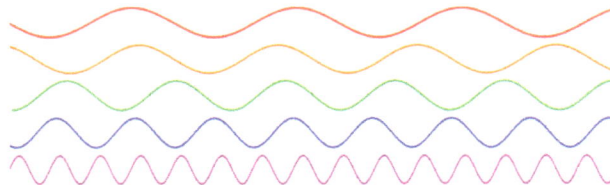

Figure 2. Sinusoidal waves of various frequencies; the bottom waves have higher frequencies than those above. The horizontal axis represents time.

However, in order to understand the sound more fully, a complex wave such as this is usually separated into its component parts, which are a combination of various sound wave frequencies (and noise). Figure 2 shows an example of a series of component sound waves such as might be seen if the clarinet sound wave was broken down into its component sine waves, but with the lower frequency components removed (the frequency ratios shown in figure 2 are too close together to be low frequency components of a sound).

Sound waves are often simplified to a description in terms of sinusoidal plane waves, which are characterized by these generic properties:

- Frequency, or its inverse, the Wavelength
- Amplitude
- Sound pressure / Intensity
- Speed of sound
- Direction

Sound that is perceptible by humans has frequencies from about 20 Hz to 20,000 Hz. In air at standard temperature and pressure, the corresponding wavelengths of sound waves range from

17 m to 17 mm. Sometimes speed and direction are combined as a velocity vector; wave number and direction are combined as a wave vector.

Transverse waves, also known as shear waves, have the additional property, *polarization*, and are not a characteristic of sound waves.

Speed of Sound

The speed of sound depends on the medium that the waves pass through, and is a fundamental property of the material. The first significant effort towards the measure of the speed of sound was made by Newton. He believed that the speed of sound in a particular substance was equal to the square root of the pressure acting on it (STP) divided by its density.

U.S. Navy F/A-18 approaching the sound barrier. The white halo is formed by condensed water droplets thought to result from a drop in air pressure around the aircraft.

$$c = \sqrt{\frac{p}{\rho}}$$

This was later proven wrong when found to incorrectly derive the speed. French mathematician Laplace corrected the formula by deducing that the phenomenon of sound travelling is not isothermal, as believed by Newton, but adiabatic. He added another factor to the equation-*gamma*-and multiplied $\sqrt{\gamma}$ to $\sqrt{\frac{p}{\rho}}$, thus coming up with the equation $c = \sqrt{\gamma \cdot \frac{p}{\rho}}$. Since $K = \gamma \cdot p$ the final equation came up to be $c = \sqrt{\frac{K}{\rho}}$ which is also known as the Newton-Laplace equation. In this equation, K = elastic bulk modulus, c = velocity of sound, and ρ = density. Thus, the speed of sound is proportional to the square root of the ratio of the bulk modulus of the medium to its density.

Those physical properties and the speed of sound change with ambient conditions. For example, the speed of sound in gases depends on temperature. In 20 °C (68 °F) air at sea level, the speed of sound is approximately 343 m/s (1,230 km/h; 767 mph) using the formula "v = (331 + 0.6 T) m/s". In fresh water, also at 20 °C, the speed of sound is approximately 1,482 m/s (5,335 km/h; 3,315 mph). In steel, the speed of sound is about 5,960 m/s (21,460 km/h; 13,330 mph). The speed of sound is also slightly sensitive, being subject to a second-order anharmonic effect, to the sound amplitude, which means that there are non-linear propagation effects, such as the production of harmonics and mixed tones not present in the original sound.

Perception of Sound

A distinct use of the term *sound* from its use in physics is that in physiology and psychology, where the term refers to the subject of *perception* by the brain. The field of psychoacoustics is dedicated to such studies. Historically the word "sound" referred exclusively to an effect in the mind. Webster's 1947 dictionary defined sound as: "that which is heard; the effect which is produced by the vibration of a body affecting the ear." This meant (at least in 1947) the correct response to the question: "if a tree falls in the forest with no one to hear it fall, does it make a sound?" was "no". However, owing to contemporary usage, definitions of sound as a physical effect are prevalent in most dictionaries. Consequently, the answer to the same question is now "yes, a tree falling in the forest with no one to hear it fall does make a sound".

The physical reception of sound in any hearing organism is limited to a range of frequencies. Humans normally hear sound frequencies between approximately 20 Hz and 20,000 Hz (20 kHz), The upper limit decreases with age. Sometimes *sound* refers to only those vibrations with frequencies that are within the hearing range for humans or sometimes it relates to a particular animal. Other species have different ranges of hearing. For example, dogs can perceive vibrations higher than 20 kHz, but are deaf below 40 Hz.

As a signal perceived by one of the major senses, sound is used by many species for detecting danger, navigation, predation, and communication. Earth's atmosphere, water, and virtually any physical phenomenon, such as fire, rain, wind, surf, or earthquake, produces (and is characterized by) its unique sounds. Many species, such as frogs, birds, marine and terrestrial mammals, have also developed special organs to produce sound. In some species, these produce song and speech. Furthermore, humans have developed culture and technology (such as music, telephone and radio) that allows them to generate, record, transmit, and broadcast sound.

Elements of Sound Perception

There are six experimentally separable ways in which sound waves are analysed. They are: pitch, duration, loudness, timbre, sonic texture and spatial location.

Figure. Pitch perception

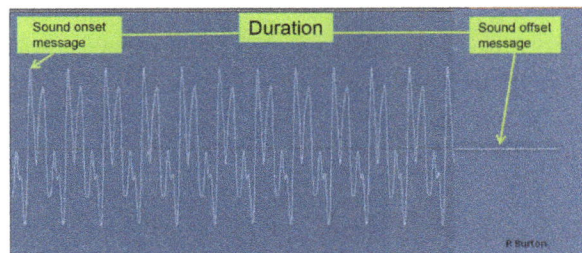

Figure. Duration perception

Pitch

Pitch is perceived as how "low" or "high" a sound is and represents the cyclic, repetitive nature of the vibrations that make up sound. For simple sounds, pitch relates to the frequency of the slowest vibration in the sound (called the fundamental harmonic). In the case of complex sounds, pitch perception can vary. Sometimes individuals identify different pitches for the same sound, based on their personal experience of particular sound patterns. Selection of a particular pitch is determined by pre-conscious examination of vibrations, including their frequencies and the balance between them. Specific attention is given to recognising potential harmonics. Every sound is placed on a pitch continuum from low to high. For example: white noise (random noise spread evenly across all frequencies) sounds higher in pitch than pink noise (random noise spread evenly across octaves) as white noise has more high frequency content. Figure shows an example of pitch recognition. During the listening process, each sound is analysed for a repeating pattern and the results for-warded to the auditory cortex as a single pitch of a certain height (octave) and chroma (note name).

Duration

Duration is perceived as how "long" or "short" a sound is and relates to onset and offset signals created by nerve responses to sounds. The duration of a sound usually lasts from the time the sound is first noticed until the sound is identified as having changed or ceased. Sometimes this is not directly related to the physical duration of a sound. For example; in a noisy environment, gapped sounds (sounds that stop and start) can sound as if they are continuous because the offset messages are missed owing to disruptions from noises in the same general bandwidth. This can be of great benefit in understanding distorted messages such as radio signals that suffer from in-terference, as (owing to this effect) the message is heard as if it was continuous. Figure gives an example of duration identification. When a new sound is noticed, a sound onset message is sent to the auditory cortex. When the repeating pattern is missed, a sound offset messages is sent.

Loudness

Loudness is perceived as how "loud" or "soft" a sound is and relates to the totalled number of auditory nerve stimulations over short cyclic time periods, most likely over the duration of theta wave cycles. This means that at short durations, a very short sound can sound softer than a longer sound even though they are presented at the same intensity level. Past around 200 ms this is no longer the case and the duration of the sound no longer affects the apparent loudness of the sound. Figure 3 gives an impression of how loudness information is summed over a period of about 200 ms before being sent to the auditory cortex. Louder signals create a greater 'push' on the Basilar membrane and thus stimulate more nerves,creating a stronger loudness signal. A more complex signal also creates more nerve firings and so sounds louder (for the same wave amplitude) than a simpler sound, such as a sine wave.

Timbre

Timbre is perceived as the quality of different sounds (e.g. the thud of a fallen rock, the whir of a drill, the tone of a musical instrument or the quality of a voice) and represents the pre-conscious allocation of a sonic identity to a sound (e.g. "it's an oboe!"). This identity is based on information gained from frequency transients, noisiness, unsteadiness, perceived pitch and the spread and intensity of overtones in the sound over an extended time frame. The way a sound changes over time provides most of the information for timbre identification. Even though a small section of the wave form from each instrument looks very similar, differences in changes over time between the clarinet and the piano are evident in both loudness and harmonic content. Less noticeable are the different noises heard, such as air hisses for the clarinet and hammer strikes for the piano.

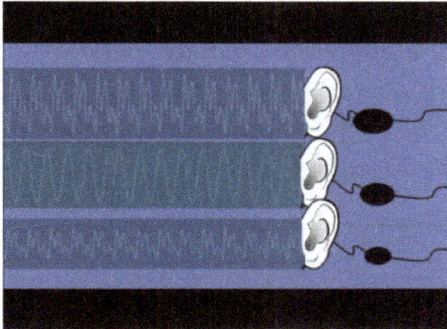

Figure 3. Loudness perception

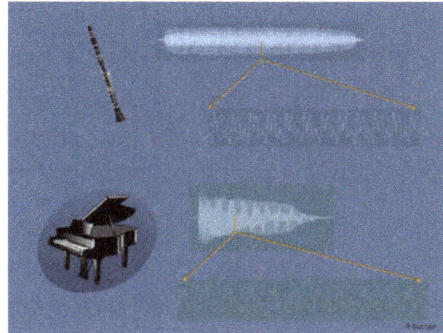

Figure 4. Timbre perception

Sonic Texture

Sonic texture relates to the number of sound sources and the interaction between them. The word 'texture', in this context, relates to the cognitive separation of auditory objects. In music, texture is often referred to as the difference between unison, polyphony and homophony, but it can also relate (for example) to a busy cafe; a sound which might be referred to as 'cacophony'. However texture refers to more than this. The texture of an orchestral piece is very different to the texture of a brass quartet because of the different numbers of players. The texture of a market place is very different to a school hall because of the differences in the various sound sources.

Spatial Location

Spatial location represents the cognitive placement of a sound in an environmental context; including the placement of a sound on both the horizontal and vertical plane, the distance from the sound source and the characteristics of the sonic environment. In a thick texture, it is possible to identify multiple sound sources using a combination of spatial location and timbre identification. It is the main reason why we can pick the sound of an oboe in an orchestra and the words of a single person at a cocktail party.

Noise

Noise is a term often used to refer to an unwanted sound. In science and engineering, noise is an undesirable component that obscures a wanted signal. However, in sound perception it can often be used to identify the source of a sound and is an important component of timbre perception.

Soundscape

Soundscape is the component of the acoustic environment that can be perceived by humans. The acoustic environment is the combination of all sounds (whether audible to humans or not) within a given area as modified by the environment.

Sound Pressure Level

Sound pressure is the difference, in a given medium, between average local pressure and the pressure in the sound wave. A square of this difference (i.e., a square of the deviation from the equilibrium pressure) is usually averaged over time and/or space, and a square root of this average provides a root mean square (RMS) value. For example, 1 Pa RMS sound pressure (94 dBSPL) in atmospheric air implies that the actual pressure in the sound wave oscillates between (1 atm $-\sqrt{2}$ Pa) and (1 atm $+\sqrt{2}$ Pa), that is between 101323.6 and 101326.4 Pa. As the human ear can detect sounds with a wide range of amplitudes, sound pressure is often measured as a level on a logarithmic decibel scale. The sound pressure level (SPL) or L_p is defined as

$$L_p = 10\log_{10}\left(\frac{p^2}{p_{ref}^2}\right) = 20\log_{10}\left(\frac{p}{p_{ref}}\right) \text{ dB}$$

where p is the root-mean-square sound pressure and p_{ref} is a reference sound pressure. Commonly used reference sound pressures, defined in the standard ANSI S1.1-1994, are 20 µPa in air and 1 µPa in water. Without a specified reference sound pressure, a value expressed in decibels cannot represent a sound pressure level.

Since the human ear does not have a flat spectral response, sound pressures are often frequency weighted so that the measured level matches perceived levels more closely. The International Electrotechnical Commission (IEC) has defined several weighting schemes. A-weighting attempts to match the response of the human ear to noise and A-weighted sound pressure levels are labeled dBA. C-weighting is used to measure peak levels.

Ultrasound

Ultrasounds are sound waves with frequencies higher than the upper audible limit of human hearing. Ultrasound is no different from 'normal' (audible) sound in its physical properties, except in that humans cannot hear it. This limit varies from person to person and is approximately 20 kilohertz (20,000 hertz) in healthy, young adults. Ultrasound devices operate with frequencies from 20 kHz up to several gigahertz.

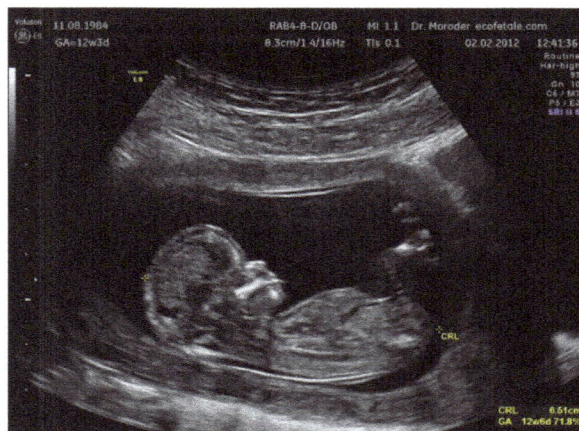

Ultrasound image of a fetus in the womb, viewed at 12 weeks of pregnancy (bidimensional-scan)

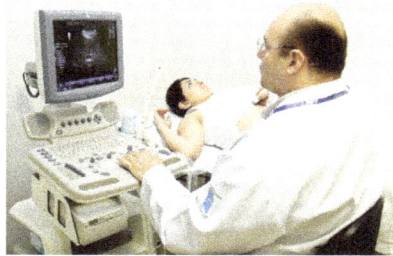
An ultrasonic examination

Ultrasound is used in many different fields. Ultrasonic devices are used to detect objects and measure distances. Ultrasound imaging or sonography is often used in medicine. In the nondestructive testing of products and structures, ultrasound is used to detect invisible flaws. Industrially, ultrasound is used for cleaning, mixing, and to accelerate chemical processes. Animals such as bats and porpoises use ultrasound for locating prey and obstacles. Scientist are also studying ultrasound using graphene diaphragms as a method of communication.

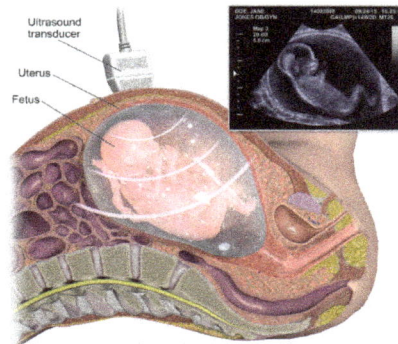
Fetal Ultrasound
Fetal ultrasound

History

Acoustics, the science of sound, starts as far back as Pythagoras in the 6th century BC, who wrote on the mathematical properties of stringed instruments. Sir Francis Galton constructed a whistle producing ultrasound in 1893. The first technological application of ultrasound was an attempt to detect submarines by Paul Langevin in 1917. The piezoelectric effect, discovered by Jacques and Pierre Curie in 1880, was useful in transducers to generate and detect ultrasonic waves in air and water. Echolocation in bats was discovered by Lazzaro Spallanzani in 1794, when he demonstrated that bats hunted and navigated by inaudible sound and not vision.

Definition

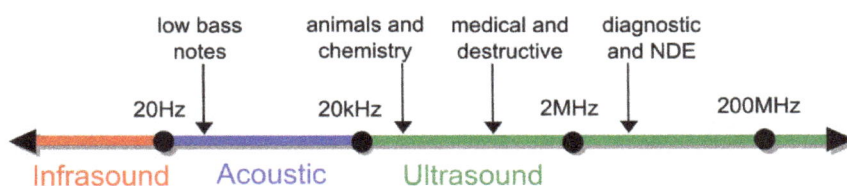

Approximate frequency ranges corresponding to ultrasound, with rough guide of some applications

Ultrasound is defined by the American National Standards Institute as "sound at frequencies greater than 20 kHz."

Perception

Humans

The upper frequency limit in humans (approximately 20 kHz) is due to limitations of the middle ear. Auditory sensation can occur if high-intensity ultrasound is fed directly into the human skull and reaches the cochlea through bone conduction, without passing through the middle ear.

Children can hear some high-pitched sounds that older adults cannot hear, because in humans the upper limit pitch of hearing tends to decrease with age. An American cell phone company has used this to create ring signals supposedly only able to be heard by younger humans; but many older people can hear the signals, which may be because of the considerable variation of age-related deterioration in the upper hearing threshold. The Mosquito is an electronic device that uses a high pitched frequency to deter loitering by young people.

Animals

Bats use a variety of ultrasonic ranging (echolocation) techniques to detect their prey. They can detect frequencies beyond 100 kHz, possibly up to 200 kHz.

Bats use ultrasounds to navigate in the darkness.

Many insects have good ultrasonic hearing and most of these are nocturnal insects listening for echolocating bats. This includes many groups of moths, beetles, praying mantids and lacewings. Upon hearing a bat, some insects will make evasive manoeuvres to escape being caught. Ultrasonic frequencies trigger a reflex action in the noctuid moth that cause it to drop slightly in its flight to evade attack. Tiger moths also emit clicks which may disturb bats' echolocation, but may also in other cases evade being eaten by advertising the fact that they are poisonous by emitting sound.

Dogs with normal hearing can hear ultrasound. A dog whistle exploits this by emitting a high frequency sound to call to a dog. Many such whistles emit sound in the upper audible range of humans, but some, such as the silent whistle, emit ultrasound at a frequency in the range 18–22 kHz.

Toothed whales, including dolphins, can hear ultrasound and use such sounds in their navigational system (biosonar) to orient and capture prey. Porpoises have the highest known upper hearing limit, at around 160 kHz. Several types of fish can detect ultrasound. In the order Clupeiformes, members of the subfamily Alosinae (shad), have been shown to be able to detect sounds up to 180 kHz, while the other subfamilies (e.g. herrings) can hear only up to 4 kHz.

Ultrasound generator/speaker systems are sold as electronic pest control devices, which are claimed to frighten away rodents and insects, but there is no scientific evidence that the devices work.

Detection and Ranging

Non-contact Sensor

An ultrasonic level or sensing system requires no contact with the target. For many processes in the medical, pharmaceutical, military and general industries this is an advantage over inline sensors that may contaminate the liquids inside a vessel or tube or that may be clogged by the product.

Both continuous wave and pulsed systems are used. The principle behind a pulsed-ultrasonic technology is that the transmit signal consists of short bursts of ultrasonic energy. After each burst, the electronics looks for a return signal within a small window of time corresponding to the time it takes for the energy to pass through the vessel. Only a signal received during this window will qualify for additional signal processing.

A popular consumer application of ultrasonic ranging was the Polaroid SX-70 camera which included a light-weight transducer system to focus the camera automatically. Polaroid later licensed this ultrasound technology and it became the basis of a variety of ultrasonic products.

Motion Sensors and Flow Measurement

A common ultrasound application is an automatic door opener, where an ultrasonic sensor detects a person's approach and opens the door. Ultrasonic sensors are also used to detect intruders; the ultrasound can cover a wide area from a single point. The flow in pipes or open channels can be measured by ultrasonic flowmeters, which measure the average velocity of flowing liquid. In rheology, an acoustic rheometer relies on the principle of ultrasound. In fluid mechanics, fluid flow can be measured using an ultrasonic flow meter.

Non-destructive Testing

Ultrasonic testing is a type of nondestructive testing commonly used to find flaws in materials and to measure the thickness of objects. Frequencies of 2 to 10 MHz are common but for special purposes other frequencies are used. Inspection may be manual or automated and is an essential part of modern manufacturing processes. Most metals can be inspected as well as plastics and aerospace composites. Lower frequency ultrasound (50–500 kHz) can also be used to inspect less dense materials such as wood, concrete and cement.

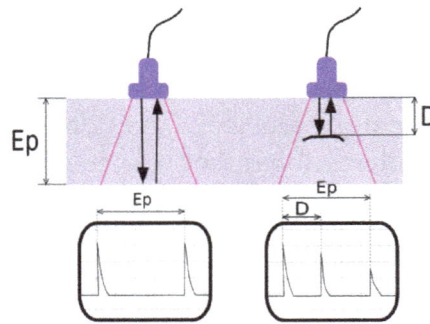

Principle of flaw detection with ultrasound. A void in the solid material reflects some energy back to the transducer, which is detected and displayed.

Ultrasound inspection of welded joints has been an alternative to radiography for non-destructive testing since the 1960s. Ultrasonic inspection eliminates the use of ionizing radiation, with safety and cost benefits. Ultrasound can also provide additional information such as the depth of flaws in a welded joint. Ultrasonic inspection has progressed from manual methods to computerized systems that automate much of the process. An ultrasonic test of a joint can identify the existence of flaws, measure their size, and identify their location. Not all welded materials are equally amenable to ultrasonic inspection; some materials have a large grain size that produces a high level of background noise in measurements.

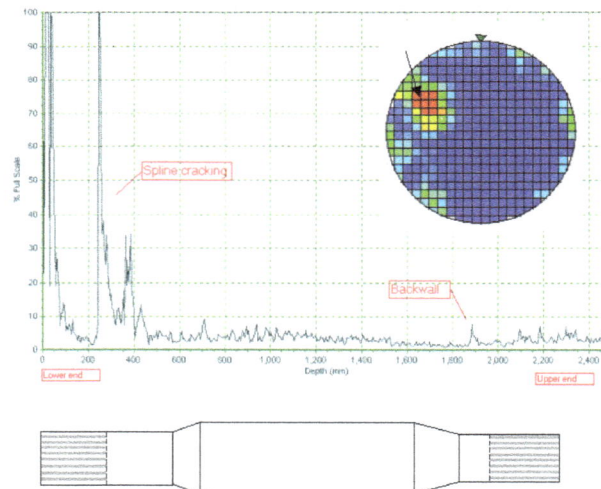

Non-destructive testing of a swing shaft showing spline cracking

Ultrasonic thickness measurement is one technique used to monitor quality of welds.

Ultrasonic Range Finding

A common use of ultrasound is in underwater range finding; this use is also called Sonar. An ultrasonic pulse is generated in a particular direction. If there is an object in the path of this pulse, part or all of the pulse will be reflected back to the transmitter as an echo and can be detected through the receiver path. By measuring the difference in time between the pulse being transmitted and the echo being received, it is possible to determine the distance.

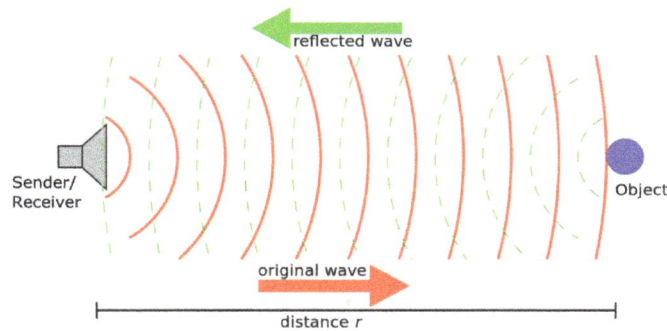

Principle of an active sonar

The measured travel time of Sonar pulses in water is strongly dependent on the temperature and the salinity of the water. Ultrasonic ranging is also applied for measurement in air and for short distances. For example, hand-held ultrasonic measuring tools can rapidly measure the layout of rooms.

Although range finding underwater is performed at both sub-audible and audible frequencies for great distances (1 to several kilometers), ultrasonic range finding is used when distances are shorter and the accuracy of the distance measurement is desired to be finer. Ultrasonic measurements may be limited through barrier layers with large salinity, temperature or vortex differentials. Ranging in water varies from about hundreds to thousands of meters, but can be performed with centimeters to meters accuracy

Ultrasound Identification (USID)

Ultrasound Identification (USID) is a Real Time Locating System (RTLS) or Indoor Positioning System (IPS) technology used to automatically track and identify the location of objects in real time using simple, inexpensive nodes (badges/tags) attached to or embedded in objects and devices, which then transmit an ultrasound signal to communicate their location to microphone sensors.

Imaging

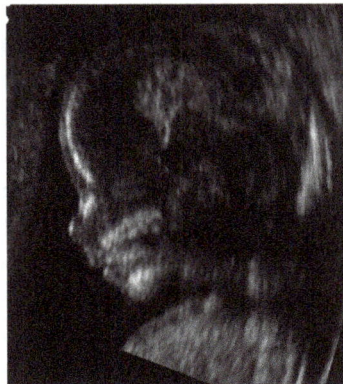

Sonogram of a fetus at 14 weeks (profile)

The potential for ultrasonic imaging of objects, with a 3 GHZ sound wave producing resolution comparable to an optical image, was recognized by Sokolov in 1939 but techniques of the time produced relatively low-contrast images with poor sensitivity. Ultrasonic imaging uses frequencies of 2 megahertz and higher; the shorter wavelength allows resolution of small internal details in structures and tissues. The power density is generally less than 1 watt per square centimetre, to avoid heating and cavitation effects in the object under examination. High and ultra high ultrasound waves are used in acoustic microscopy, with frequencies up to 4 gigahertz. Ultrasonic imaging applications include industrial non-destructive testing, quality control and medical uses.

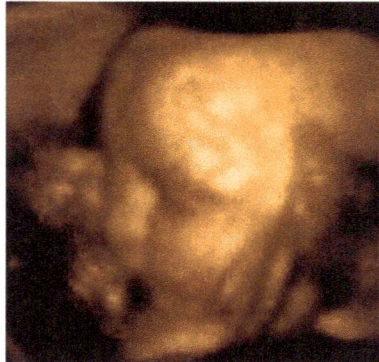

Head of a fetus, aged 29 weeks, in a "3D ultrasound"

Acoustic Microscopy

Acoustic microscopy is the technique of using sound waves to visualize structures too small to be resolved by the human eye. Frequencies up to several gigahertz are used in acoustic microscopes. The reflection and diffraction of sound waves from microscopic structures can yield information not available with light.

Human Medicine

Medical sonography (ultrasonography) is an ultrasound-based diagnostic medical imaging technique used to visualize muscles, tendons, and many internal organs, to capture their size, structure and any pathological lesions with real time tomographic images. Ultrasound has been used by radiologists and sonographers to image the human body for at least 50 years and has become a widely used diagnostic tool. The technology is relatively inexpensive and portable, especially when compared with other techniques, such as magnetic resonance imaging (MRI) and computed tomography (CT). Ultrasound is also used to visualize fetuses during routine and emergency prenatal care. Such diagnostic applications used during pregnancy are referred to as obstetric sonography. As currently applied in the medical field, properly performed ultrasound poses no known risks to the patient. Sonography does not use ionizing radiation, and the power levels used for imaging are too low to cause adverse heating or pressure effects in tissue. Although the long-term effects due to ultrasound exposure at diagnostic intensity are still unknown, currently most doctors feel that the benefits to patients outweigh the risks. The ALARA (As Low As Reasonably Achievable) principle has been advocated for an ultrasound examination – that is, keeping the scanning time and power settings as low as possible but consistent with diagnostic imaging – and that by that principle non-medical uses, which by definition are not necessary, are actively discouraged.

Ultrasound is also increasingly being used in trauma and first aid cases, with emergency ultrasound becoming a staple of most EMT response teams. Furthermore, ultrasound is used in remote diagnosis cases where teleconsultation is required, such as scientific experiments in space or mobile sports team diagnosis.

According to RadiologyInfo, ultrasounds are useful in the detection of pelvic abnormalities and can involve techniques known as abdominal (transabdominal) ultrasound, vaginal (transvaginal or endovaginal) ultrasound in women, and also rectal (transrectal) ultrasound in men.

Veterinary Medicine

Diagnostic ultrasound is used externally in horses for evaluation of soft tissue and tendon injuries, and internally in particular for reproductive work – evaluation of the reproductive tract of the mare and pregnancy detection. It may also be used in an external manner in stallions for evaluation of testicular condition and diameter as well as internally for reproductive evaluation (deferent duct etc.).

Starting at the turn of the century, ultrasound technology began to be used by the beef cattle industry to improve animal health and the yield of cattle operations. Ultrasound is used to evaluate fat thickness, rib eye area, and intramuscular fat in living animals. It is also used to evaluate the health and characteristics of unborn calves.

Ultrasound technology provides a means for cattle producers to obtain information that can be used to improve the breeding and husbandry of cattle. The technology can be expensive, and it requires a substantial time commitment for continuous data collection and operator training. Nevertheless, this technology has proven useful in managing and running a cattle breeding operation.

Processing and Power

High-power applications of ultrasound often use frequencies between 20 kHz and a few hundred kHz. Intensities can be very high; above 10 watts per square centimeter, cavitation can be inducted in liquid media, and some applications use up to 1000 watts per square centimeter. Such high intensities can induce chemical changes or produce significant effects by direct mechanical action, and can inactivate harmful microorganisms.

Physical Therapy

Ultrasound has been used since the 1940s by physical and occupational therapists for treating connective tissue: ligaments, tendons, and fascia (and also scar tissue). Conditions for which ultrasound may be used for treatment include the follow examples: ligament sprains, muscle strains, tendonitis, joint inflammation, plantar fasciitis, metatarsalgia, facet irritation, impingement syndrome, bursitis, rheumatoid arthritis, osteoarthritis, and scar tissue adhesion.

Biomedical Applications

Ultrasound also has therapeutic applications, which can be highly beneficial when used with dosage precautions Relatively high power ultrasound can break up stony deposits or tissue, accelerate the effect of drugs in a targeted area, assist in the measurement of the elastic properties of tissue, and can be used to sort cells or small particles for research.

Ultrasonic Impact Treatment

Ultrasonic impact treatment (UIT) uses ultrasound to enhance the mechanical and physical properties of metals. It is a metallurgical processing technique in which ultrasonic energy is applied to a metal object. Ultrasonic treatment can result in controlled residual compressive stress, grain refinement and grain size reduction. Low and high cycle fatigue are enhanced and have been documented to provide increases up to ten times greater than non-UIT specimens. Additionally, UIT has proven effective in addressing stress corrosion cracking, corrosion fatigue and related issues.

When the UIT tool, made up of the ultrasonic transducer, pins and other components, comes into contact with the work piece it acoustically couples with the work piece, creating harmonic resonance. This harmonic resonance is performed at a carefully calibrated frequency, to which metals respond very favorably.

Depending on the desired effects of treatment a combination of different frequencies and displacement amplitude is applied. These frequencies range between 25 and 55 kHz, with the displacement amplitude of the resonant body of between 22 and 50 μm (0.00087 and 0.0020 in).

UIT Devices Rely on Magnetostrictive Transducers.

Processing

Ultrasonication offers great potential in the processing of liquids and slurries, by improving the mixing and chemical reactions in various applications and industries. Ultrasonication generates alternating low-pressure and high-pressure waves in liquids, leading to the formation and violent collapse of small vacuum bubbles. This phenomenon is termed cavitation and causes high speed impinging liquid jets and strong hydrodynamic shear-forces. These effects are used for the deagglomeration and milling of micrometre and nanometre-size materials as well as for the disintegration of cells or the mixing of reactants. In this aspect, ultrasonication is an alternative to high-speed mixers and agitator bead mills. Ultrasonic foils under the moving wire in a paper machine will use the shock waves from the imploding bubbles to distribute the cellulose fibres more uniformly in the produced paper web, which will make a stronger paper with more even surfaces. Furthermore, chemical reactions benefit from the free radicals created by the cavitation as well as from the energy input and the material transfer through boundary layers. For many processes, this sonochemical effect leads to a substantial reduction in the reaction time, like in the transesterification of oil into biodiesel.

Schematic of bench and industrial-scale ultrasonic liquid processors

Substantial ultrasonic intensity and high ultrasonic vibration amplitudes are required for many processing applications, such as nano-crystallization, nano-emulsification, deagglomeration, extraction, cell disruption, as well as many others. Commonly, a process is first tested on a laboratory scale to prove feasibility and establish some of the required ultrasonic exposure parameters. After this phase is complete, the process is transferred to a pilot (bench) scale for flow-through pre-production optimization and then to an industrial scale for continuous production. During these scale-up steps, it is essential to make sure that all local exposure conditions (ultrasonic amplitude, cavitation intensity, time spent in the active cavitation zone, etc.) stay the same. If this condition is met, the quality of the final product remains at the optimized level, while the productivity is increased by a predictable "scale-up factor". The productivity increase results from the fact that laboratory, bench and industrial-scale ultrasonic processor systems incorporate progressively larger ultrasonic horns, able to generate progressively larger high-intensity cavitation zones and, therefore, to process more material per unit of time. This is called "direct scalability". It is important to point out that increasing the power of the ultrasonic processor alone does *not* result in direct scalability, since it may be (and frequently is) accompanied by a reduction in the ultrasonic amplitude and cavitation intensity. During direct scale-up, all processing conditions must be maintained, while the power rating of the equipment is increased in order to enable the operation of a larger ultrasonic horn.

Ultrasonic Manipulation and Characterization of Particles

A researcher at the Industrial Materials Research Institute, Alessandro Malutta, devised an experiment that demonstrated the trapping action of ultrasonic standing waves on wood pulp fibers diluted in water and their parallel orienting into the equidistant pressure planes. The time to orient the fibers in equidistant planes is measured with a laser and an electro-optical sensor. This could provide the paper industry a quick on-line fiber size measurement system. A somewhat different implementation was demonstrated at Pennsylvania State University using a microchip which generated a pair of perpendicular standing surface acoustic waves allowing to position particles equidistant to each other on a grid. This experiment, called "acoustic tweezers", can be used for applications in material sciences, biology, physics, chemistry and nanotechnology.

Ultrasonic Cleaning

Ultrasonic cleaners, sometimes mistakenly called *supersonic cleaners*, are used at frequencies from 20 to 40 kHz for jewellery, lenses and other optical parts, watches, dental instruments, surgical instruments, diving regulators and industrial parts. An ultrasonic cleaner works mostly by energy released from the collapse of millions of microscopic cavitations near the dirty surface. The bubbles made by cavitation collapse forming tiny jets directed at the surface.

Ultrasonic Disintegration

Similar to ultrasonic cleaning, biological cells including bacteria can be disintegrated. High power ultrasound produces cavitation that facilitates particle disintegration or reactions. This has uses in biological science for analytical or chemical purposes (sonication and sonoporation) and in killing bacteria in sewage. High power ultrasound can disintegrate corn slurry and enhance liquefaction and saccharification for higher ethanol yield in dry corn milling plants.

Ultrasonic Humidifier

The ultrasonic humidifier, one type of nebulizer (a device that creates a very fine spray), is a popular type of humidifier. It works by vibrating a metal plate at ultrasonic frequencies to nebulize (sometimes incorrectly called "atomize") the water. Because the water is not heated for evaporation, it produces a cool mist. The ultrasonic pressure waves nebulize not only the water but also materials in the water including calcium, other minerals, viruses, fungi, bacteria, and other impurities. Illness caused by impurities that reside in a humidifier's reservoir fall under the heading of "Humidifier Fever".

Ultrasonic humidifiers are frequently used in aeroponics, where they are generally referred to as foggers.

Ultrasonic Welding

In ultrasonic welding of plastics, high frequency (15 kHz to 40 kHz) low amplitude vibration is used to create heat by way of friction between the materials to be joined. The interface of the two parts is specially designed to concentrate the energy for maximum weld strength.

Sonochemistry

Power ultrasound in the 20–100 kHz range is used in chemistry. The ultrasound does not interact directly with molecules to induce the chemical change, as its typical wavelength (in the millimeter range) is too long compared to the molecules. Instead, the energy causes cavitation which generates extremes of temperature and pressure in the liquid where the reaction happens. Ultrasound also breaks up solids and removes passivating layers of inert material to give a larger surface area for the reaction to occur over. Both of these effects make the reaction faster. In 2008, Atul Kumar reported synthesis of Hantzsch esters and polyhydroquinoline derivatives via multi-component reaction protocol in aqueous micelles using ultrasound.

Ultrasound is used in extraction, using different frequencies.

Weapons

Ultrasound has been studied as a basis for sonic weapons, for applications such as riot control, disorientation of attackers, up to lethal levels of sound.

Wireless Communication

In July 2015, The Economist reported that researchers at the University of California, Berkeley have conducted ultrasound studies using graphene diaphragms. The thinness and low weight of graphene combined with its strength make it an effective material to use with for ultrasound communications. One suggested application of the technology would be underwater communications, where radio waves typically do not travel well.

Other Uses

Ultrasound when applied in specific configurations can produce short bursts of light in an exotic phenomenon known as sonoluminescence. This phenomenon is being investigated partly because of the possibility of bubble fusion (a nuclear fusion reaction hypothesized to occur during sonoluminescence).

Ultrasound is used when characterizing particulates through the technique of ultrasound attenuation spectroscopy or by observing electroacoustic phenomena or by transcranial pulsed ultrasound.

Audio can be propagated by modulated ultrasound.

A formerly popular consumer application of ultrasound was in television remote controls for adjusting volume and changing channels. Introduced by Zenith in the late 1950s, the system used a hand-held remote control containing short rod resonators struck by small hammers, and a microphone on the set. Filters and detectors discriminated between the various operations. The principal advantages were that no battery was needed in the hand-held control box, and unlike radio waves, the ultrasound was unlikely to affect neighboring sets. Ultrasound remained in use until displaced by infrared systems starting in the late 1980s.

Safety

Occupational exposure to ultrasound in excess of 120 dB may lead to hearing loss. Exposure in excess of 155 dB may produce heating effects that are harmful to the human body, and it has been calculated that exposures above 180 dB may lead to death. The UK's independent Advisory Group on Non-ionising Radiation (AGNIR) produced a report in 2010, which was published by the UK Health Protection Agency (HPA). This report recommended an exposure limit for the general public to airborne ultrasound sound pressure levels (SPL) of 70 dB (at 20 kHz), and 100 dB (at 25 kHz and above).

Infrasound

Infrasound, sometimes referred to as low-frequency sound, is sound that is lower in frequency than 20 Hz (hertz) or cycles per second, the "normal" limit of human hearing. Hearing becomes gradually less sensitive as frequency decreases, so for humans to perceive infrasound, the sound pressure must be sufficiently high. The ear is the primary organ for sensing infrasound, but at higher intensities it is possible to feel infrasound vibrations in various parts of the body.

Infrasound arrays at infrasound monitoring station in Qaanaaq, Greenland.

The study of such sound waves is referred to sometimes as **infrasonics**, covering sounds beneath 20 Hz down to 0.1Hz and rarely to 0.001 Hz. This frequency range is utilized for monitoring earthquakes, charting rock and petroleum formations below the earth, and also in ballistocardiography and seismocardiography to study the mechanics of the heart.

Infrasound is characterized by an ability to cover long distances and get around obstacles with little dissipation. In music, low frequency sounds, including near infrasound, can be produced using acoustic waveguide methods, such as a large pipe organ or, for reproduction, exotic loudspeaker designs such as transmission line, rotary woofer, or traditional subwoofer designs. Subwoofers designed to produce infrasound are capable of sound reproduction an octave or more below that of most commercially available subwoofers, and are often about 10 times the size.

History and Study

Infrasound was used by the Allies of World War I to locate artillery. One of the pioneers in infrasonic research was French scientist Vladimir Gavreau. His interest in infrasonic waves first came about in his laboratory during the 1960s, when he and his laboratory assistants experienced pain in the ear drums and shaking laboratory equipment, but no audible sound was picked up on his microphones. He concluded it was infrasound caused by a large fan and duct system and soon got to work preparing tests in the laboratories. One of his experiments was an infrasonic whistle, an oversized organ pipe.

Definition

Infrasound is defined by the American National Standards Institute as "sound at frequencies less than 20 Hz."

Sources

Infrasound can result from both natural and human sources:

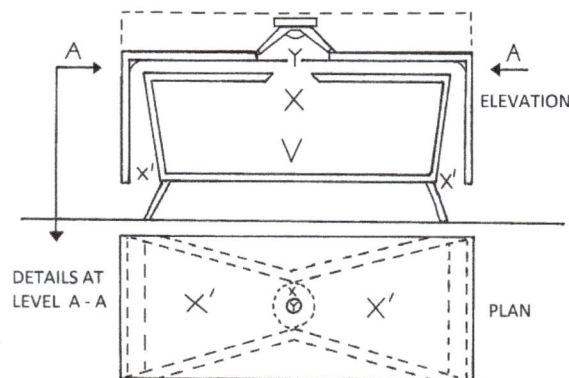

Patent for a double bass reflex loudspeaker enclosure design intended to produce infrasonic frequencies ranging from 5 to 25 hertz, of which traditional subwoofer designs are not readily capable.

- Natural events: infrasonic sound sometimes results naturally from severe weather, surf, lee waves, avalanches, earthquakes, volcanoes, bolides, waterfalls, calving of icebergs, aurorae, meteors, lightning and upper-atmospheric lightning. Nonlinear ocean wave interactions in ocean storms produce pervasive infrasound vibrations around 0.2 Hz, known as microbaroms. According to the Infrasonics Program at NOAA, infrasonic arrays can be used to locate avalanches in the Rocky Mountains, and to detect tornadoes on the high plains several minutes before they touch down.

- Animal communication: whales, elephants, hippopotamuses, rhinoceros, giraffes, okapi, and alligators are known to use infrasound to communicate over distances—up to hundreds of miles in the case of whales. In particular, the Sumatran Rhinoceros has been shown to produce sounds with frequencies as low as 3 Hz which have similarities with the song of the humpback whale. The roar of the tiger contains infrasound of 18 Hz and lower, and the purr of felines is reported to cover a range of 20 to 50 Hz. It has also been suggested that migrating birds use naturally generated infrasound, from sources such as turbulent airflow over mountain ranges, as a navigational aid. Infrasound also may be used for long-distance communication, especially well documented in baleen whales and African elephants. The frequency of baleen whale sounds can range from 10 Hz to 31 kHz, and that of elephant calls from 15 Hz to 35 Hz. Both can be extremely loud (around 117 dB), allowing communication for many kilometres, with a possible maximum range of around 10 km (6 mi) for elephants, and potentially hundreds or thousands of kilometers for some whales. Elephants also produce infrasound waves that travel through solid ground and are sensed by other herds using their feet, although they may be separated by hundreds of kilometres. These calls may be used to coordinate the movement of herds and allow mating elephants to find each other.

- Human singers: some vocalists, including Tim Storms, can produce notes in the infrasound range.

- Human created sources: infrasound can be generated by human processes such as sonic booms and explosions (both chemical and nuclear), or by machinery such as diesel engines, wind turbines and specially designed mechanical transducers (industrial vibration tables). Certain specialized loudspeaker designs are also able to reproduce extremely low frequencies; these include large-scale rotary woofer models of subwoofer loudspeaker, as well as large horn loaded, bass reflex, sealed and transmission line loudspeakers.

Animal Reactions

Animals have been known to perceive the infrasonic waves going through the earth by natural disasters and can use these as an early warning. A recent example of this is the 2004 Indian Ocean earthquake and tsunami. Animals were reported to flee the area hours before the actual tsunami hit the shores of Asia. It is not known for sure if this is the exact cause, as some have suggested that it may have been the influence of electromagnetic waves, and not of infrasonic waves, that prompted these animals to flee.

Research in 2013 by Jon Hagstrum of the US Geological Survey suggests that homing pigeons use low frequency infrasound to navigate.

Human Reactions

20 Hz is considered the normal low-frequency limit of human hearing. When pure sine waves are reproduced under ideal conditions and at very high volume, a human listener will be able to identify tones as low as 12 Hz. Below 10 Hz it is possible to perceive the single cycles of the sound, along with a sensation of pressure at the eardrums.

The dynamic range of the auditory system decreases with decreasing frequency. This compression can be seen in the equal-loudness-level contours, and it implies that a slight increase in level can change the perceived loudness from barely audible to loud. Combined with the natural spread in thresholds within a population, its effect may be that a very low-frequency sound which is inaudible to some people may be loud to others.

One study has suggested that infrasound may cause feelings of awe or fear in humans. It has also been suggested that since it is not consciously perceived, it may make people feel vaguely that odd or supernatural events are taking place.

A scientist working at Sydney University's Auditory Neuroscience Laboratory reports growing evidence that infrasound may affect some people's nervous system by stimulating the vestibular system and this has shown in animal models an effect similar to sea sickness. In a study of 45 people, Tehran University researchers stated: "Despite all the good benefits of wind turbines ... this technology has health risks for all those exposed to its sound" — in particular, sleep disorder. In another study by researchers at Ibaraki University in Japan said the EEG tests showed the brain function showed that the infrasound produced by wind turbine were "considered to be an annoyance to the technicians who work in close to a modern large-scale wind turbine".

Infrasonic 17 Hz Tone Experiment

On 31 May 2003, a group of UK researchers held a mass experiment where they exposed some 700 people to music laced with soft 17 Hz sine waves played at a level described as "near the edge of hearing", produced by an extra-long-stroke subwoofer mounted two-thirds of the way from the end of a seven-meter-long plastic sewer pipe. The experimental concert (entitled *Infrasonic*) took place in the Purcell Room over the course of two performances, each consisting of four musical pieces. Two of the pieces in each concert had 17 Hz tones played underneath.

In the second concert, the pieces that were to carry a 17 Hz undertone were swapped so that test results would not focus on any specific musical piece. The participants were not told which pieces included the low-level 17 Hz near-infrasonic tone. The presence of the tone resulted in a significant number (22%) of respondents reporting anxiety, uneasiness, extreme sorrow, nervous feelings of revulsion or fear, chills down the spine, and feelings of pressure on the chest.

In presenting the evidence to the British Association for the Advancement of Science, Professor Richard Wiseman said, "These results suggest that low frequency sound can cause people to have unusual experiences even though they cannot consciously detect infrasound. Some scientists have suggested that this level of sound may be present at some allegedly haunted sites and so cause people to have odd sensations that they attribute to a ghost—our findings support these ideas."

Suggested Relationship to Ghost Sightings

Psychologist Richard Wiseman of the University of Hertfordshire suggests that the odd sensations that people attribute to ghosts may be caused by infrasonic vibrations. In 1998, Vic Tandy, experimental officer and part-time lecturer in the school of international studies and law at Coventry University, and Dr. Tony Lawrence of the psychology department wrote a paper called "Ghosts in the Machine" for the *Journal of the Society for Psychical Research*. Their research suggested that an infrasonic signal of 19 Hz might be responsible for some ghost sightings. Tandy was working late one night alone in a supposedly haunted laboratory at Warwick, when he felt very anxious and could detect a grey blob out of the corner of his eye. When Tandy turned to face the grey blob, there was nothing.

The following day, Tandy was working on his fencing foil, with the handle held in a vice. Although there was nothing touching it, the blade started to vibrate wildly. Further investigation led Tandy to discover that the extractor fan in the lab was emitting a frequency of 18.98 Hz, very close to the resonant frequency of the eye given as 18 Hz by NASA. This was why Tandy conjectured that he had seen a ghostly figure— he believed that it was an optical illusion caused by his eyeballs resonating. The room was exactly half a wavelength in length, and the desk was in the centre, thus causing a standing wave which caused the vibration of the foil.

Tandy investigated this phenomenon further and wrote a paper entitled *The Ghost in the Machine*. Tandy carried out a number of investigations at various sites believed to be haunted, including the basement of the Tourist Information Bureau next to Coventry Cathedral and Edinburgh Castle.

Detection and Measurement

NASA Langley has designed and developed an infrasonic detection system which can be used to make useful infrasound measurements at a location where it was not possible previously. The system comprises an electret condenser microphone PCB Model 377M06, having a 3-inch membrane diameter, and a small, compact windscreen. Electret-based technology offers the lowest possible background noise, because Johnson noise generated in the supporting electronics (preamplifier) is minimized.

The microphone features a high membrane compliance with a large backchamber volume, a prepolarized backplane and a high impedance preamplifier located inside the backchamber. The windscreen, based on the high transmission coefficient of infrasound through matter, is made of a material having a low acoustic impedance and sufficiently thick wall to insure structural stability. Close-cell polyurethane foam has been found to serve the purpose well. In the proposed test, test parameters will be sensitivity, background noise, signal fidelity (harmonic distortion), and temporal stability.

The microphone design differs from that of a conventional audio system in that the peculiar features of infrasound are taken into account. First, infrasound propagates over vast distances through the Earth's atmosphere as a result of very low atmospheric absorption and refractive ducting that enables propagation by way of multiple bounces between the Earth's surface and the stratosphere. A second property that has received little attention is the great penetration capability of infrasound through solid matter – a property utilized in the design and fabrication of the system windscreens.

Thus the system fulfills several instrumentation requirements advantageous to the application of acoustics: (1) a low frequency microphone with especially low background noise, which enables detection of low-level signals within a low-frequency passband; (2) a small, compact windscreen that permits (3) rapid deployment of a microphone array in the field. The system also features a data acquisition system that permits real time detection, bearing, and signature of a low-frequency source.

The Comprehensive Nuclear-Test-Ban Treaty Organization Preparatory Commission uses infrasound as one of its monitoring technologies, along with seismic, hydroacoustic, and atmospheric radionuclide monitoring. The loudest infrasound recorded to date by the monitoring system was generated by the 2013 Chelyabinsk meteor.

References

- Maia, Silva. Theoretical And Experimental Modal Analysis, Research Studies Press Ltd., 1997, ISBN 0-471-97067-0.

- Novelline, Robert (1997). Squire's Fundamentals of Radiology (5th ed.). Harvard University Press. pp. 34–35. ISBN 0-674-83339-2.

- Bruno Pollet Power Ultrasound in Electrochemistry: From Versatile Laboratory Tool to Engineering Solution John Wiley & Sons, 2012 ISBN 1119967864,

- Whitlow W. L. Au (1993). The sonar of dolphins. Springer. ISBN 978-0-387-97835-2. Retrieved 13 November 2011.

- Emmanuel P. Papadakis (ed) Ultrasonic Instruments & Devices, Academic Press, 1999 ISBN 0-12-531951-7 page 752

- Gail D Betts et al, Inactivation of Food-borne Microrganisms using Power Ultrasound in Encyclopedia of Food Microbiology, Academic Press, 2000 ISBN 0-12-227070-3 page 2202

- Part II, industrial; commercial applications (1991). Guidelines for the Safe Use of Ultrasound Part II – Industrial & Commercial Applications – Safety Code 24. Health Canada. ISBN 0-660-13741-0.

- E. von Muggenthaler, C. Baes, D. Hill, R. Fulk, A. Lee: Infrasound and low frequency vocalizations from the giraffe; Helmholtz resonance in biology, proceedings of Riverbanks Consortium on biology and behavior, 1999. Also work by Muggenthaler et al cited by Nicole Herget: Giraffes, Living Wild, Creative Education, 2009, ISBN 978-1-58341-654-9, p. 38

- Richardson, Greene, Malme, Thomson (1995). Marine Mammals and Noise. Academic Press. ISBN 978-0-12-588440-2.

- Hsu, Christine (24 August 2012). "Man With World's Deepest Voice Hits Notes That Only Elephants Can Hear". Medical Daily. Retrieved 2 August 2016. American singer Tim Storms who also has the world's widest vocal range can reach notes as low as G-7 (0.189Hz) [...] so low that even Storms himself cannot hear it.

- "Effect of Wind Turbine Noise on Workers' Sleep Disorder: A Case Study of Manjil Wind Farm in Northern Iran". Tehran University of Medical Sciences. Retrieved 29 April 2015.

- Burton, R. L. (2015). The elements of music: what are they, and who cares? In J. Rosevear & S. Harding. (Eds.), ASME XXth National Conference proceedings. Paper presented at: Music: Educating for life: ASME XXth National Conference (pp.22 - 28), Parkville, Victoria: The Australian Society for Music Education Inc.

- American Institute for Ultrasound in Medicine practice guidelines (linked PDFs). American Institute for Ultrasound in Medicine. Retrieved 2015-07-01

- Development and installation of an infrasonic wake vortex detection system By Qamar A. Shams and Allan J. Zuckerwar, NASA Langley Research Center, Hampton VA USA, WakeNet-Europe 2014, Bretigny, France.

- NASA Langley Researchers Nab Invention of the Year for Infrasound Detection System By Joe Atkinson, 2014, NASA Langley Research Center

- Paul Harper (20 February 2013). "Meteor explosion largest infrasound recorded". The New Zealand Herald. APN Holdings NZ. Retrieved 31 March 2013.

5

Transduction in Acoustics

A device that converts one form of energy to another is a transducer. This chapter explains to the reader the importance of transducers in acoustics. There are several transduction devices in use on day-to-day basis, like loudspeakers, microphones and hydrophones. These devices have been explained in the following chapter.

Transducer

A transducer is a device that converts one form of energy to another. Usually a transducer converts a signal in one form of energy to a signal in another.

Transducers are often employed at the boundaries of automation, measurement, and control systems, where electrical signals are converted to and from other physical quantities (energy, force, torque, light, motion, position, etc.). The process of converting one form of energy to another is known as transduction.

Transducer Types

Passive

Passive sensors require an external power sources to operate, which is called an excitation signal. The signal is modulated by the sensor to produce the output signal. For example, a thermistor does not generate any electric signal, but by passing electric current through it, its resistance can be measured by detecting variations in current and/or voltage across the thermistor.

Passive

Passive sensors generate electric signals in response to an external stimulus without the need of an additional energy source. Such examples are a thermocouple, photodiode, and a piezoelectric sensor.

Sensors

A sensor is a device that receives and responds to a signal or stimulus. Transducer is the other term that is sometimes interchangeably used instead of the term sensor, although there are subtle differences. A transducer is a term that can be used for the definition of many devices such as sensors, actuators, or transistors.

Actuators

An actuator is a device that is responsible for moving or controlling a mechanism or system. It is operated by a source of energy, which can be mechanical force, electrical current, hydraulic fluid pressure, or pneumatic pressure, and converts that energy into motion. An actuator is the mechanism by which a control system acts upon an environment. The control system can be simple (a fixed mechanical or electronic system), software-based (e.g. a printer driver, robot control system), a human, or any other input.

Bidirectional

Bidirectional transducers convert physical phenomena to electrical signals and also convert electrical signals into physical phenomena. Examples of inherently bidirectional transducers are antennae, which can convert conducted electrical signals to or from propagating electromagnetic waves, and voice coils, which convert electrical signals into sound (when used in a loudspeaker) or sound into electrical signals (when used in a microphone). Likewise, DC electric motors may be used to generate electrical power if the motor shaft is turned by an external torque.

Loudspeaker

A loudspeaker (or loud-speaker or speaker) is an electroacoustic transducer; which converts an electrical audio signal into a corresponding sound. The first primitive loudspeakers were invented during the development of telephone systems in the late 1800s, but electronic amplification by vacuum tube beginning around 1912 made loudspeakers truly practical. By the 1920s they were used in radios, phonographs, public address systems and theatre sound systems for talking motion pictures.

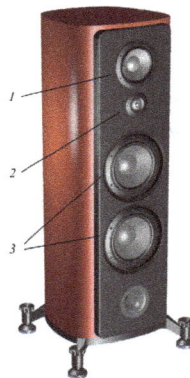

Loudspeaker for home use with three types of dynamic drivers
1. Mid-range driver
2. Tweeter
3. Woofers

The most widely used type of speaker today is the dynamic speaker, invented in 1925 by Edward W. Kellogg and Chester W. Rice. The dynamic speaker operates on the same basic principle as a dynamic microphone, but in reverse, to produce sound from an electrical signal. When an alternating current electrical audio signal is applied to its voice coil, a coil of wire suspended in a circular gap between the poles of a permanent magnet, the coil is forced to move rapidly back and forth due to Faraday's law of induction, which causes a diaphragm (usually conically shaped) attached to the coil to move back and forth, pushing on the air to create sound waves. Besides this most common method, there are several alternative technologies that can be used to convert an electrical signal into sound. The sound source (e.g., a sound recording or a microphone) must be amplified with an amplifier before the signal is sent to the speaker.

Speakers are typically housed in an enclosure which is often a rectangular or square box made of wood or sometimes plastic, and the enclosure plays an important role in the quality of the sound. Where high fidelity reproduction of sound is required, multiple loudspeaker transducers are often mounted in the same enclosure, each reproducing a part of the audible frequency range *(picture at right)*. In this case the individual speakers are referred to as "drivers" and the entire unit is called a loudspeaker. Drivers made for reproducing high audio frequencies are called tweeters, those for middle frequencies are called mid-range drivers, and those for low frequencies are called woofers. Smaller loudspeakers are found in devices such as radios, televisions, portable audio players, computers, and electronic musical instruments . Larger loudspeaker systems are used for music, sound reinforcement in theatres and concerts, and in public address systems.

Terminology

The term "loudspeaker" may refer to individual transducers (known as "drivers") or to complete speaker systems consisting of an enclosure including one or more drivers.

To adequately reproduce a wide range of frequencies with even coverage, most loudspeaker systems employ more than one driver, particularly for higher sound pressure level or maximum accuracy. Individual drivers are used to reproduce different frequency ranges. The drivers are named subwoofers (for very low frequencies); woofers (low frequencies); mid-range speakers (middle frequencies); tweeters (high frequencies); and sometimes supertweeters, optimized for the highest audible frequencies. The terms for different speaker drivers differ, depending on the application. In two-way systems there is no mid-range driver, so the task of reproducing the mid-range sounds falls upon the woofer and tweeter. Home stereos use the designation "tweeter" for the high frequency driver, while professional concert systems may designate them as "HF" or "highs". When multiple drivers are used in a system, a "filter network", called a crossover, separates the incoming signal into different frequency ranges and routes them to the appropriate driver. A loudspeaker system with n separate frequency bands is described as "n-way speakers": a two-way system will have a woofer and a tweeter; a three-way system employs a woofer, a mid-range, and a tweeter. Loudspeaker driver of the type pictured are termed "dynamic" (short for electrodynamic) to distinguish them from earlier drivers (i.e., moving iron speaker), or speakers using piezoelectric or electrostatic systems, or any of several other sorts.

History

Johann Philipp Reis installed an electric loudspeaker in his *telephone* in 1861; it was capable of reproducing clear tones, but also could reproduce muffled speech after a few revisions. Alexander Graham Bell patented his first electric loudspeaker (capable of reproducing intelligible speech) as part of his telephone in 1876, which was followed in 1877 by an improved version from Ernst Siemens. During this time, Thomas Edison was issued a British patent for a system using compressed air as an amplifying mechanism for his early cylinder phonographs, but he ultimately settled for the familiar metal horn driven by a membrane attached to the stylus. In 1898, Horace Short patented a design for a loudspeaker driven by compressed air; he then sold the rights to Charles Parsons, who was issued several additional British patents before 1910. A few companies, including the Victor Talking Machine Company and Pathé, produced record players using compressed-air loudspeakers. However, these designs were significantly limited by their poor sound quality and their inability to reproduce sound at low volume. Variants of the system were used for public address applications, and more recently, other variations have been used to test space-equipment resistance to the very loud sound and vibration levels that the launching of rockets produces.

The first experimental moving-coil (also called *dynamic*) loudspeaker was invented by Oliver Lodge in 1898. The first practical moving-coil loudspeakers were manufactured by Danish engineer Peter L. Jensen and Edwin Pridham in 1915, in Napa, California. Like previous loudspeakers these used horns to amplify the sound produced by a small diaphragm. Jensen was denied patents. Being unsuccessful in selling their product to telephone companies, in 1915 they changed their target market to radios and public address systems, and named their product Magnavox. Jensen was, for years after the invention of the loudspeaker, a part owner of The Magnavox Company.

Kellogg and Rice in 1925 holding the large driver of the first moving-coil cone loudspeaker.

Prototype moving-coil cone loudspeaker by Kellogg and Rice in 1925, with electromagnet pulled back, showing voice coil attached to cone

The first commercial version of the speaker, sold with the RCA Radiola receiver, had only a 6 inch cone. In 1926 it sold for $250, equivalent to about $3000 today.

The moving-coil principle commonly used today in speakers was patented in 1924 by Chester W. Rice and Edward W. Kellogg. The key difference between previous attempts and the patent by Rice and Kellogg is the adjustment of mechanical parameters so that the fundamental resonance of the moving system is below the frequency where the cone's radiation impedance becomes uniform.

About this same period, Walter H. Schottky invented the first ribbon loudspeaker together with Dr. Erwin Gerlach.

These first loudspeakers used electromagnets, because large, powerful permanent magnets were generally not available at a reasonable price. The coil of an electromagnet, called a field coil, was energized by current through a second pair of connections to the driver. This winding usually served a dual role, acting also as a choke coil, filtering the power supply of the amplifier that the loudspeaker was connected to. AC ripple in the current was attenuated by the action of passing through the choke coil. However, AC line frequencies tended to modulate the audio signal going to the voice coil and added to the audible hum. In 1930 Jensen introduced the first commercial fixed-magnet loudspeaker; however, the large, heavy iron magnets of the day were impractical and field-coil speakers remained predominant until the widespread availability of lightweight Alnico magnets after World War II.

In the 1930s, loudspeaker manufacturers began to combine two and three bandpasses' worth of drivers in order to increase frequency response and sound pressure level. In 1937, the first film industry-standard loudspeaker system, "The Shearer Horn System for Theatres" (a two-way system), was introduced by Metro-Goldwyn-Mayer. It used four 15″ low-frequency drivers, a crossover network set for 375 Hz, and a single multi-cellular horn with two compression drivers providing the high frequencies. John Kenneth Hilliard, James Bullough Lansing, and Douglas Shearer all played roles in creating the system. At the 1939 New York World's Fair, a very large two-way public address system was mounted on a tower at Flushing Meadows. The eight 27″ low-frequency drivers were designed by Rudy Bozak in his role as chief engineer for Cinaudagraph. High-frequency drivers were likely made by Western Electric.

Altec Lansing introduced the *604*, which became their most famous coaxial Duplex driver, in 1943. It incorporated a high-frequency horn that sent sound through the middle of a 15-inch woofer for near-point-source performance. Altec's "Voice of the Theatre" loudspeaker system arrived in the marketplace in 1945, offering better coherence and clarity at the high output levels necessary in movie theaters. The Academy of Motion Picture Arts and Sciences immediately began testing its sonic characteristics; they made it the film house industry standard in 1955. In 1954, Edgar Villchur developed the acoustic suspension principle of loudspeaker design in Cambridge, Massachusetts. This allowed for better bass response than previously from drivers mounted in smaller cabinets which was important during the transition to stereo recording and reproduction. He and his partner Henry Kloss formed the Acoustic Research company to manufacture and market speaker systems using this principle. Subsequently, continuous developments in enclosure design and materials led to significant audible improvements. The most notable improvements to date in modern dynamic drivers, and the loudspeakers that employ them, are improvements in cone materials, the introduction of higher-temperature adhesives, improved permanent magnet materials, improved measurement techniques, computer-aided design, and finite element analysis. At low frequencies, the application of electrical network theory to the acoustic performance allowed by various enclosure designs (initially by Thiele, and later by Small) has been very important at the design level.

Driver Design - Dynamic Loudspeakers

The most common type of driver, commonly called a dynamic loudspeaker, uses a lightweight diaphragm, or *cone*, connected to a rigid *basket*, or *frame*, via a flexible suspension, commonly called a *spider*, that constrains a voice coil to move axially through a cylindrical magnetic gap. When an electrical signal is applied to the voice coil, a magnetic field is created by the electric current in the voice coil, making it a variable electromagnet. The coil and the driver's magnetic system interact, generating a mechanical force that causes the coil (and thus, the attached cone) to move back and forth, thereby reproducing sound under the control of the applied electrical signal coming from the amplifier. The following is a description of the individual components of this type of loudspeaker.

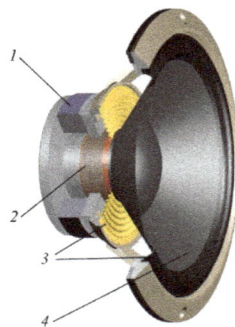

Cutaway view of a dynamic loudspeaker for the bass register.
1. Magnet
2. Voicecoil
3. Suspension
4. Diaphragm

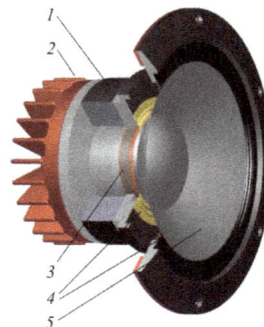

Cutaway view of a dynamic midrange speaker.
1. Magnet
2. Cooler (sometimes present)
3. Voicecoil
4. Suspension
5. Membrane

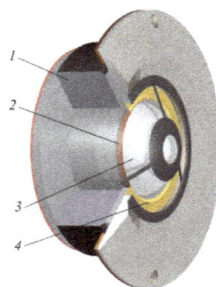

Cutaway view of a dynamic tweeter with acoustic lens and a dome-shaped membrane.
1. Magnet
2. Voicecoil
3. Membrane
4. Suspension

The diaphragm is usually manufactured with a cone- or dome-shaped profile. A variety of different materials may be used, but the most common are paper, plastic, and metal. The ideal material would 1) be rigid, to prevent uncontrolled cone motions; 2) have low mass, to minimize starting force requirements and energy storage issues; 3) be well damped, to reduce vibrations continuing after the signal has stopped with little or no audible ringing due to its resonance frequency as determined by its usage. In practice, all three of these criteria cannot be met simultaneously using existing materials; thus, driver design involves trade-offs. For example, paper is light and typically well damped, but is not stiff; metal may be stiff and light, but it usually has poor damping; plastic can be light, but typically, the stiffer it is made, the poorer the damping. As a result, many cones are made of some sort of composite material. For example, a cone might be made of cellulose paper, into which some carbon fiber, Kevlar, glass, hemp or bamboo fibers have been added; or it might use a honeycomb sandwich construction; or a coating might be applied to it so as to provide additional stiffening or damping.

The chassis, frame, or basket, is designed to be rigid, avoiding deformation that could change critical alignments with the magnet gap, perhaps causing the voice coil to rub against the sides of the gap. Chassis are typically cast from aluminum alloy, or stamped from thin steel sheet, though in some drivers with large magnets cast chassis are preferable seeing as sheet metal can easily be warped in whenever the loudspeaker is subjected to rough handling. Other materials such as molded plastic and damped plastic compound baskets are becoming common, especially for inexpensive, low-mass drivers. Metallic chassis can play an important role in conducting heat away from the voice coil; heating during operation changes resistance, causes physical dimensional changes, and if extreme, may even demagnetize permanent magnets.

The suspension system keeps the coil centered in the gap and provides a restoring (centering) force that returns the cone to a neutral position after moving. A typical suspension system consists of two parts: the *spider*, which connects the diaphragm or voice coil to the frame and provides the majority of the restoring force, and the *surround*, which helps center the coil/cone assembly and allows free pistonic motion aligned with the magnetic gap. The spider is usually made of a corrugated fabric disk, impregnated with a stiffening resin. The name comes from the shape of early suspensions, which were two concentric rings of Bakelite material, joined by six or eight curved "legs." Variations of this topology included the addition of a felt disc to provide a barrier to particles that might otherwise cause the voice coil to rub. The German firm Rulik still offers drivers with uncommon spiders made of wood.

The cone surround can be rubber or polyester foam, or a ring of corrugated, resin coated fabric; it is attached to both the outer diaphragm circumference and to the frame. These different surround materials, their shape and treatment can dramatically affect the acoustic output of a driver; each class and implementation having advantages and disadvantages. Polyester foam, for example, is lightweight and economical, but is degraded by exposure to ozone, UV light, humidity and elevated temperatures, significantly limiting useful life with adequate performance.

The wire in a voice coil is usually made of copper, though aluminum—and, rarely, silver—may be used. The advantage of aluminum is its light weight, which raises the resonant frequency of the voice coil and allows it to respond more easily to higher frequencies. A disadvantage of aluminum is that it is not easily soldered, and so connections are instead often crimped together and sealed. These connections must be made well or they may fail in an intense environment of mechanical vibration. Voice-coil wire cross sections can be circular, rectangular, or hexagonal, giving varying amounts of wire volume coverage in the magnetic gap space. The coil is oriented co-axially inside the gap; it moves back and forth within a small circular volume (a hole, slot, or groove) in the magnetic structure. The gap establishes a concentrated magnetic field between the two poles of a permanent magnet; the outside of the gap being one pole, and the center post (called the pole piece) being the other. The pole piece and backplate are often a single piece, called the poleplate or yoke.

Modern driver magnets are almost always permanent and made of ceramic, ferrite, Alnico, or, more recently, rare earth such as neodymium and samarium cobalt. A trend in design—due to increases in transportation costs and a desire for smaller, lighter devices (as in many home theater multi-speaker installations)—is the use of the last instead of heavier ferrite types. Very few manufacturers still produce electrodynamic loudspeakers with electrically powered field coils, as was common in the earliest designs; one of the last is a French firm. When high field-strength permanent magnets became available, Alnico, an alloy of aluminum, nickel, and cobalt became popular, since it dispensed with the power supply problems of field-coil drivers. Alnico was used almost exclusively until about 1980. Alnico magnets can be partially degaussed (i.e., demagnetized) by accidental 'pops' or 'clicks' caused by loose connections, especially if used with a high power amplifier. This damage can be reversed by "recharging" the magnet.

After 1980, most (but not quite all) driver manufacturers switched from Alnico to ferrite magnets, which are made from a mix of ceramic clay and fine particles of barium or strontium ferrite. Although the energy per kilogram of these ceramic magnets is lower than Alnico, it is substantially less expensive, allowing designers to use larger yet more economical magnets to achieve a given performance.

The size and type of magnet and details of the magnetic circuit differ, depending on design goals. For instance, the shape of the pole piece affects the magnetic interaction between the voice coil and the magnetic field, and is sometimes used to modify a driver's behavior. A "shorting ring", or Faraday loop, may be included as a thin copper cap fitted over the pole tip or as a heavy ring situated within the magnet-pole cavity. The benefits of this complication is reduced impedance at high frequencies, providing extended treble output, reduced harmonic distortion, and a reduction in the inductance modulation that typically accompanies large voice coil excursions. On the other hand, the copper cap requires a wider voice-coil gap, with increased magnetic reluctance; this reduces available flux, requiring a larger magnet for equivalent performance.

Driver design—including the particular way two or more drivers are combined in an enclosure to make a speaker system—is both an art and science. Adjusting a design to improve performance is done using a combination of magnetic, acoustic, mechanical, electrical, and material science theory, and tracked with high precision measurements and the observations of experienced listeners. A few of the issues speaker and driver designers must confront are distortion, radiation lobing, phase effects, off-axis response, and crossover artifacts. Designers can use an anechoic chamber to ensure the speaker can be measured independently of room effects, or any of several electronic techniques that, to some extent, substitute for such chambers. Some developers eschew anechoic chambers in favor of specific standardized room setups intended to simulate real-life listening conditions.

Fabrication of finished loudspeaker systems has become segmented, depending largely on price, shipping costs, and weight limitations. High-end speaker systems, which are typically heavier (and often larger) than economic shipping allows outside local regions, are usually made in their target market region and can cost $140,000 or more per pair. Economical mass market speaker systems and drivers may be manufactured in China or other low-cost manufacturing locations.

Driver Types

Individual electrodynamic drivers provide their best performance within a limited frequency range. Multiple drivers (e.g., subwoofers, woofers, mid-range drivers, and tweeters) are generally combined into a complete loudspeaker system to provide performance beyond that constraint.

A four-way, high fidelity loudspeaker system. Each of the four drivers outputs a different frequency range; the fifth aperture at the bottom is a bass reflex port.

The three most commonly used sound radiation systems are the cone, dome and horn type drivers.

Full-range Drivers

A full-range driver is a speaker designed to be used alone to reproduce an audio channel without the help of other drivers, and therefore must cover the entire audio frequency range. These drivers are small, typically 3 to 8 inches (7.6 to 20.3 cm) in diameter to permit reasonable high frequency response, and carefully designed to give low-distortion output at low frequencies, though with reduced maximum output level. Full-range (or more accurately, wide-range) drivers are most commonly heard in public address systems, in televisions (although some models are suitable for hi-fi listening), small radios, intercoms, some computer speakers, etc. In hi-fi speaker systems, the use of wide-range drive units can avoid undesirable interactions between multiple drivers caused by non-coincident driver location or crossover network issues. Fans of wide-range driver hi-fi speaker systems claim a coherence of sound due to the single source and a resulting lack of interference, and likely also to the lack of crossover components. Detractors typically cite wide-range drivers' limited frequency response and modest output abilities (most especially at low frequencies), together with their requirement for large, elaborate, expensive enclosures—such as transmission lines, quarter wave resonators or horns—to approach optimum performance. With the advent of neodymium drivers, low cost quarter wave transmission lines are made possible and are increasingly made availably commercially.

Full-range drivers often employ an additional cone called a *whizzer*: a small, light cone attached to the joint between the voice coil and the primary cone. The whizzer cone extends the high-frequency response of the driver and broadens its high frequency directivity, which would otherwise be greatly narrowed due to the outer diameter cone material failing to keep up with the central voice coil at higher frequencies. The main cone in a whizzer design is manufactured so as to flex more in the outer diameter than in the center. The result is that the main cone delivers low frequencies and the whizzer cone contributes most of the higher frequencies. Since the whizzer cone is smaller than the main diaphragm, output dispersion at high frequencies is improved relative to an equivalent single larger diaphragm.

Limited-range drivers, also used alone, are typically found in computers, toys, and clock radios. These drivers are less elaborate and less expensive than wide-range drivers, and they may be severely compromised to fit into very small mounting locations. In these applications, sound quality is a low priority. The human ear is remarkably tolerant of poor sound quality, and the distortion inherent in limited-range drivers may enhance their output at high frequencies, increasing clarity when listening to spoken word material.

Subwoofer

A subwoofer is a woofer driver used only for the lowest part of the audio spectrum: typically below 200 Hz for consumer systems, below 100 Hz for professional live sound, and below 80 Hz in THX-approved systems. Because the intended range of frequencies is limited, subwoofer system design is usually simpler in many respects than for conventional loudspeakers, often consisting of a single driver enclosed in a suitable box or enclosure. Since sound in this frequency range can easily bend around corners by diffraction, the speaker aperture does not have to face the audience, and subwoofers can be mounted in the bottom of the enclosure, facing the floor. This eased by the limitations of human hearing at low frequencies; such sounds cannot be located in space, due to their large wavelengths compared to higher frequencies which produce differential effects in the ears due to shadowing by the head, and diffraction around it, both of which we rely upon for localization clues.

To accurately reproduce very low bass notes without unwanted resonances (typically from cabinet panels), subwoofer systems must be solidly constructed and properly braced; good subwoofers are typically quite heavy. Many subwoofer systems include power amplifiers and electronic subsonic (sub)-filters, with additional controls relevant to low-frequency reproduction. These variants are known as "active" or "powered" subwoofers. In contrast, "passive" subwoofers require external amplification. In typical installations, subwoofers are physically separated from the rest of the transducers. Because of propagation delay, their output may be somewhat out of phase from another subwoofer (on another channel). is slightly out of phase with the rest of the sound. Consequently, a subwoofer's power amp often has a phase-delay adjustment (approximately 1 ms of delay is required for each additional foot of separation from the listener) which may improve performance of the system as a whole at subwoofer frequencies (and perhaps an octave or so above the crossover point). However, the influence of room resonances (sometimes called standing waves) is typically so large that such issues are secondary in practice.

Woofer

A woofer is a driver that reproduces low frequencies. The driver works with the characteristics of the enclosure to produce suitable low frequencies. Indeed, both are so closely connected that they must be considered together in use. Only at design time do the separate properties of enclosure and woofer matter individually. Some loudspeaker systems use a woofer for the lowest frequencies, sometimes well enough that a subwoofer is not needed. Additionally, some loudspeakers use the woofer to handle middle frequencies, eliminating the mid-range driver. This can be accomplished with the selection of a tweeter that can work low enough that, combined with a woofer that responds high enough, the two drivers add coherently in the middle frequencies.

Mid-range Driver

A mid-range speaker is a loudspeaker driver that reproduces a band of frequencies generally between 1-6Khz, otherwise known as the 'mid' frequencies (between the woofer and tweeter). Mid-range driver diaphragms can be made of paper or composite materials, and can be direct radiation drivers (rather like smaller woofers) or they can be compression drivers (rather like some tweeter designs). If the mid-range driver is a direct radiator, it can be mounted on the front baffle of a loudspeaker enclosure, or, if a compression driver, mounted at the throat of a horn for added output level and control of radiation pattern.

Tweeter

A tweeter is a high-frequency driver that reproduces the highest frequencies in a speaker system. A major problem in tweeter design is achieving wide angular sound coverage (off-axis response), since high frequency sound tends to leave the speaker in narrow beams. Soft-dome tweeters are widely found in home stereo systems, and horn-loaded compression drivers are common in professional sound reinforcement. Ribbon tweeters have gained popularity in recent years, as the output power of some designs has been increased to levels useful for professional sound reinforcement, and their output pattern is wide in the horizontal plane, a pattern that has convenient applications in concert sound.

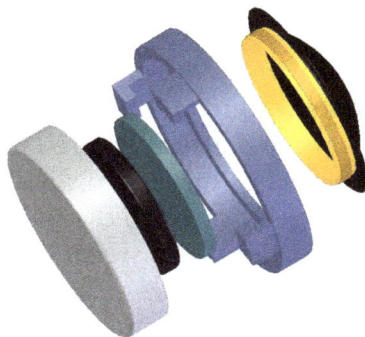

Exploded view of a dome tweeter.

Coaxial Drivers

A coaxial driver is a loudspeaker driver with two or several combined concentric drivers. Coaxial drivers have been produced by many companies, such as Altec, Tannoy, Pioneer, KEF, SEAS, B&C Speakers, BMS, Cabasse and Genelec.

Loudspeaker System Design

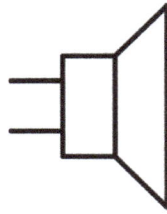

Electronic symbol for a speaker

Crossover

Used in multi-driver speaker systems, the crossover is a subsystem that separates the input signal into different frequency ranges suited to each driver. The drivers receive power only in their usable frequency range (the range they were designed for), thereby reducing distortion in the drivers and interference between them. No crossover can be perfect (i.e., absolute block at the edges of the passband, no amplitude variation within the passband, no phase changes across the frequency band boundaries the crossover establishes, ..), so this is an idealized description.

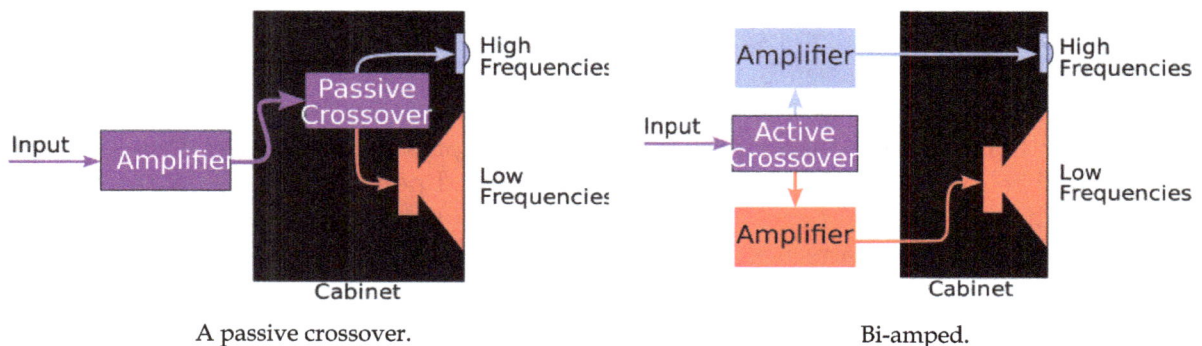

A passive crossover.

Bi-amped.

Crossovers can be *passive* or *active*. A passive crossover is an electronic circuit that uses a combination of one or more resistors, inductors, or non-polar capacitors. These parts are formed into carefully designed networks and are most often placed between the full frequency-range power amplifier and the loudspeaker drivers to divide the amplifier's signal into the necessary frequency bands before being delivered to the individual drivers. Passive crossover circuits need no external power beyond the audio signal itself, but have disadvantages: high cost, large components (inductors and capacitors), limited ability to adjust the circuit as desired due to limited choice of high power level components, etc. They also cause substantial overall signal loss and a significant reduction in damping factor between the voice coil and the crossover. An active crossover is an electronic filter circuit that divides the signal into individual frequency bands *before* power amplification, thus requiring at least one power amplifier for each bandpass. Passive filtering may also be used in this way before power amplification, but it is an uncommon solution, being less flexible than active filtering. Any technique that uses crossover filtering followed by amplification is commonly known as bi-amping, tri-amping, quad-amping, and so on, depending on the minimum number of amplifier channels. Some loudspeaker designs use a combination of passive and active crossover filtering, such as a passive crossover between the mid- and high-frequency drivers and an active crossover between the low-frequency driver and the combined mid- and high frequencies.

Passive crossovers are commonly installed inside speaker boxes and are by far the most usual type of crossover for home and low-power use. In car audio systems, passive crossovers may be in a separate box, necessary to accommodate the size of the components used. Passive crossovers may be simple for low-order filtering, or complex to allow steep slopes such as 18 or 24 dB per octave. Passive crossovers can also be designed to compensate for undesired characteristics of driver, horn, or enclosure resonances, and can be tricky to implement, due to component interaction. Passive crossovers, like the driver units that they feed, have power handling limits, have insertion losses (10% is often claimed), and change the load seen by the amplifier. The changes are matters of concern for many in the hi-fi world. When high output levels are required, active crossovers may be preferable. Active crossovers may be simple circuits that emulate the response of a passive network, or may be more complex, allowing extensive audio adjustments. Some active crossovers, usually digital loudspeaker management systems, may include facilities for precise alignment of phase and time between frequency bands, equalization, and dynamics (compression and limiting) control.

Enclosures

Most loudspeaker systems consist of drivers mounted in an enclosure, or cabinet. The role of the enclosure is to prevent sound waves emanating from the back of a driver from interfering destructively with those from the front. The sound waves emitted from the back are 180° out of phase with those emitted forward, so without an enclosure they typically cause cancellations which significantly degrade the level and quality of sound at low frequencies.

The simplest driver mount is a flat panel (i.e., baffle) with the drivers mounted in holes in it. However, in this approach, sound frequencies with a wavelength longer than the baffle dimensions are canceled out, because the antiphase radiation from the rear of the cone interferes with the radiation from the front. With an infinitely large panel, this interference could be entirely prevented. A sufficiently large sealed box can approach this behavior.

An unusual three-way speaker system. The cabinet is narrow to raise the frequency where a diffraction effect called the "baffle step" occurs.

Since panels of infinite dimensions are impossible, most enclosures function by containing the rear radiation from the moving diaphragm. A sealed enclosure prevents transmission of the sound emitted from the rear of the loudspeaker by confining the sound in a rigid and airtight box. Techniques used to reduce transmission of sound through the walls of the cabinet include thicker cabinet walls, lossy wall material, internal bracing, curved cabinet walls—or more rarely, visco-elastic materials (e.g., mineral-loaded bitumen) or thin lead sheeting applied to the interior enclosure walls.

However, a rigid enclosure reflects sound internally, which can then be transmitted back through the loudspeaker diaphragm—again resulting in degradation of sound quality. This can be reduced by internal absorption using absorptive materials (often called "damping"), such as glass wool, wool, or synthetic fiber batting, within the enclosure. The internal shape of the enclosure can also be designed to reduce this by reflecting sounds away from the loudspeaker diaphragm, where they may then be absorbed.

Other enclosure types alter the rear sound radiation so it can add constructively to the output from the front of the cone. Designs that do this (including *bass reflex, passive radiator, transmission line*, etc.) are often used to extend the effective low-frequency response and increase low-frequency output of the driver.

To make the transition between drivers as seamless as possible, system designers have attempted to time-align (or phase adjust) the drivers by moving one or more driver mounting locations forward or back so that the acoustic center of each driver is in the same vertical plane. This may also involve tilting the face speaker back, providing a separate enclosure mounting for each driver, or (less commonly) using electronic techniques to achieve the same effect. These attempts have resulted in some unusual cabinet designs.

The speaker mounting scheme (including cabinets) can also cause diffraction, resulting in peaks and dips in the frequency response. The problem is usually greatest at higher frequencies, where wavelengths are similar to, or smaller than, cabinet dimensions. The effect can be minimized by rounding the front edges of the cabinet, curving the cabinet itself, using a smaller or narrower enclosure, choosing a strategic driver arrangement, using absorptive material around a driver, or some combination of these and other schemes.

Horn Loudspeakers

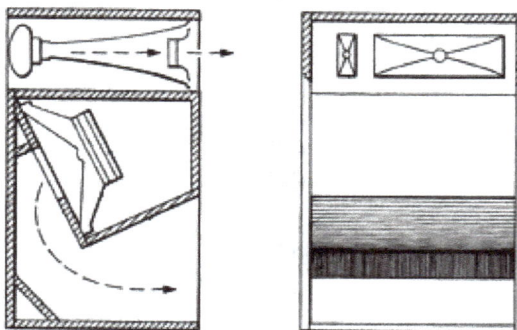

A three-way loudspeaker that uses horns in front of each of the three drivers: a shallow horn for the tweeter, a long, straight horn for mid frequencies and a folded horn for the woofer

Horn loudspeakers are the oldest form of loudspeaker system. The use of horns as voice-amplifying megaphones dates at least to the 17th century, and horns were used in mechanical gramophones as early as 1857. Horn loudspeakers use a shaped waveguide in front of or behind the driver to increase the directivity of the loudspeaker and to transform a small diameter, high pressure condition at the driver cone surface to a large diameter, low pressure condition at the mouth of the horn. This increases the sensitivity of the loudspeaker and focuses the sound over a narrower area. The size of the throat, mouth, the length of the horn, as well as the area expansion rate along it must be carefully chosen to match the drive to properly provide this transforming function over a range of frequencies (every horn performs poorly outside its acoustic limits, at both high and low frequencies). The length and cross-sectional mouth area required to create a bass or sub-bass horn require a horn many feet long. 'Folded' horns can reduce the total size, but compel designers to make compromises and accept increased complication such as cost and construction. Some horn designs not only fold the low frequency horn, but use the walls in a room corner as an extension of the horn mouth. In the late 1940s, horns whose mouths took up much of a room wall were not unknown amongst hi-fi fans. Room sized installations became much less acceptable when two or more were required.

A horn loaded speaker can have a sensitivity as high as 110 dB at 2.83 volts (1 watt at 8 ohms) at 1 meter. This is a hundredfold increase in output compared to a speaker rated at 90 dB sensitivity, and is invaluable in applications where high sound levels are required or amplifier power is limited.

Wiring Connections

Two-way binding posts on a loudspeaker, connected using banana plugs.

A 4-ohm loudspeaker with two pairs of binding posts capable of accepting bi-wiring after the removal of two metal straps.

Most loudspeakers use two wiring points to connect to the source of the signal (for example, to the audio amplifier or receiver). To accept the wire connection, the loudspeaker enclosure may have binding posts, spring clips, or a panel-mount jack. If the wires for a pair of speakers are not connected with respect to the proper electrical polarity (the + and – connections on the speaker and amplifier should be connected + to + and – to –; speaker cable is almost always marked so that one conductor of a pair can be distinguished from the other, even if it has run under or behind things in its run from amplifier to speaker location), the loudspeakers are said to be "out of phase" or more properly "out of polarity". Given identical signals, motion in one cone is in the opposite direction of the other. This typically causes monophonic material in a stereo recording to be canceled out, reduced in level, and made more difficult to localize, all due to destructive interference of the sound waves. The cancellation effect is most noticeable at frequencies where the loudspeakers are separated by a quarter wavelength or less; low frequencies are affected the most. This type of miswiring error does not damage speakers, but is not optimal for listening.

Wireless Speakers

Wireless speakers are very similar to traditional (wired) loudspeakers, but they receive audio signals using radio frequency (RF) waves rather than over audio cables. There is normally an amplifier integrated in the speaker's cabinet because the RF waves alone are not enough to drive the speaker. This integration of amplifier and loudspeaker is known as an active loudspeaker. Manufacturers of these loudspeakers design them to be as lightweight as possible while producing the maximum amount of audio output efficiency.

HP Roar Wireless Speaker

Wireless speakers still need power, so require a nearby AC power outlet, or possibly batteries. Only the wire to the amplifier is eliminated.

Specifications

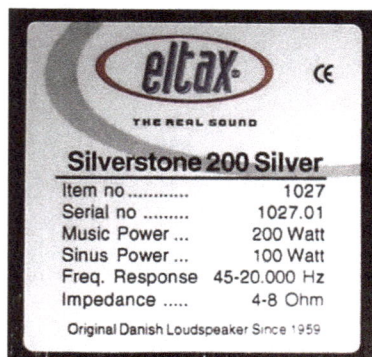

Specifications label on a loudspeaker.

Speaker specifications generally include:

- Speaker or driver type (individual units only) – Full-range, woofer, tweeter, or mid-range.

- Size of individual drivers. For cone drivers, the quoted size is generally the outside diameter of the basket. However, it may less commonly also be the diameter of the cone surround, measured apex to apex, or the distance from the center of one mounting hole to its opposite. Voice-coil diameter may also be specified. If the loudspeaker has a compression horn driver, the diameter of the horn throat may be given.

- Rated Power – Nominal (or even continuous) power, and peak (or maximum short-term) power a loudspeaker can handle (i.e., maximum input power before destroying the loudspeaker; it is never the sound output the loudspeaker produces). A driver may be damaged at much less than its rated power if driven past its mechanical limits at lower frequencies. Tweeters can also be damaged by amplifier clipping (amplifier circuits produce large amounts of energy at high frequencies in such cases) or by music or sine wave input at high frequencies. Each of these situations might pass more energy to a tweeter than it can survive without damage. In some jurisdictions, power handling has a legal meaning allowing comparisons between loudspeakers under consideration. Elsewhere, the variety of meanings for power handling capacity can be quite confusing.

- Impedance – typically $4\,\Omega$ (ohms), $8\,\Omega$, etc.

- Baffle or enclosure type (enclosed systems only) – Sealed, bass reflex, etc.

- Number of drivers (complete speaker systems only) – two-way, three-way, etc.

- and optionally:

- Crossover frequency(ies) (multi-driver systems only) – The nominal frequency boundaries of the division between drivers.

- Frequency response – The measured, or specified, output over a specified range of frequencies for a constant input level varied across those frequencies. It sometimes includes a variance limit, such as within «± 2.5 dB.»

- Thiele/Small parameters (individual drivers only) – these include the driver's F_s (resonance frequency), Q_{ts} (a driver's Q; more or less, its damping factor at resonant frequency), V_{as} (the equivalent air compliance volume of the driver), etc.

- Sensitivity – The sound pressure level produced by a loudspeaker in a non-reverberant environment, often specified in dB and measured at 1 meter with an input of 1 watt (2.83 rms volts into $8\,\Omega$), typically at one or more specified frequencies. Manufacturers often use this rating in marketing material.

- Maximum sound pressure level – The highest output the loudspeaker can manage, short of damage or not exceeding a particular distortion level. Manufacturers often use this rating in marketing material—commonly without reference to frequency range or distortion level.

Electrical Characteristics of Dynamic Loudspeakers

The load that a driver presents to an amplifier consists of a complex electrical impedance—a combination of resistance and both capacitive and inductive reactance, which combines properties of the driver, its mechanical motion, the effects of crossover components (if any are in the signal path between amplifier and driver), and the effects of air loading on the driver as modified by the enclosure and its environment. Most amplifiers' output specifications are given at a specific power into an ideal resistive load; however, a loudspeaker does not have a constant resistance across its frequency range. Instead, the voice coil is inductive, the driver has mechanical resonances, the enclosure changes the driver's electrical and mechanical characteristics, and a passive crossover between the drivers and the amplifier contributes its own variations. The result is a load resistance that varies fairly widely with frequency, and usually a varying phase relationship between voltage and current as well, also changing with frequency. Some amplifiers can cope with the variation better than others can.

To make sound, a loudspeaker is driven by modulated electric current (produced by an amplifier) that pass through a "speaker coil" which then (through inductance) magnetizes the coil, creating a magnetic field. The electric current variations that pass through the speaker are thus converted to varying magnetic forces, which move the speaker diaphragm, which thus forces the driver to produce air motion that is similar to the original signal from the amplifier.

Electromechanical Measurements

Examples of typical measurements are: amplitude and phase characteristics vs. frequency; impulse response under one or more conditions (e.g., square waves, sine wave bursts, etc.); directivity vs. frequency (e.g., horizontally, vertically, spherically, etc.); harmonic and intermodulation distortion vs. sound pressure level (SPL) output, using any of several test signals; stored energy (i.e., ringing) at various frequencies; impedance vs. frequency; and small-signal vs. large-signal performance. Most of these measurements require sophisticated and often expensive equipment to perform, and also good judgment by the operator, but the raw sound pressure level output is rather easier to report and so is often the only specified value—sometimes in misleadingly exact terms. The sound pressure level (SPL) a loudspeaker produces is measured in decibels (dB_{spl}).

Efficiency Vs. Sensitivity

Loudspeaker efficiency is defined as the sound power output divided by the electrical power input. Most loudspeakers are inefficient transducers; only about 1% of the electrical energy sent by an amplifier to a typical home loudspeaker is converted to acoustic energy. The remainder is converted to heat, mostly in the voice coil and magnet assembly. The main reason for this is the difficulty of achieving proper impedance matching between the acoustic impedance of the drive unit and the air it radiates into. (At low frequencies, improving this match is the main purpose of speaker enclosure designs). The efficiency of loudspeaker drivers varies with frequency as well. For instance, the output of a woofer driver decreases as the input frequency decreases because of the increasingly poor match between air and the driver.

Driver ratings based on the SPL for a given input are called sensitivity ratings and are notionally similar to efficiency. Sensitivity is usually defined as so many decibels at 1 W electrical input, measured at 1 meter (except for headphones), often at a single frequency. The voltage used is often 2.83 V$_{RMS}$, which is 1 watt into an 8 Ω (nominal) speaker impedance (approximately true for many speaker systems). Measurements taken with this reference are quoted as dB with 2.83 V @ 1 m.

The sound pressure output is measured at (or mathematically scaled to be equivalent to a measurement taken at) one meter from the loudspeaker and on-axis (directly in front of it), under the condition that the loudspeaker is radiating into an infinitely large space and mounted on an infinite baffle. Clearly then, sensitivity does not correlate precisely with efficiency, as it also depends on the directivity of the driver being tested and the acoustic environment in front of the actual loudspeaker. For example, a cheerleader's horn produces more sound output in the direction it is pointed by concentrating sound waves from the cheerleader in one direction, thus "focusing" them. The horn also improves impedance matching between the voice and the air, which produces more acoustic power for a given speaker power. In some cases, improved impedance matching (via careful enclosure design) lets the speaker produce more acoustic power.

- Typical home loudspeakers have sensitivities of about 85 to 95 dB for 1 W @ 1 m—an efficiency of 0.5–4%.

- Sound reinforcement and public address loudspeakers have sensitivities of perhaps 95 to 102 dB for 1 W @ 1 m—an efficiency of 4–10%.

- Rock concert, stadium PA, marine hailing, etc. speakers generally have higher sensitivities of 103 to 110 dB for 1 W @ 1 m—an efficiency of 10–20%.

A driver with a higher maximum power rating cannot necessarily be driven to louder levels than a lower-rated one, since sensitivity and power handling are largely independent properties. In the examples that follow, assume (for simplicity) that the drivers being compared have the same electrical impedance, are operated at the same frequency within both driver's respective pass bands, and that power compression and distortion are low. For the first example, a speaker 3 dB more sensitive than another produces double the sound power (is 3 dB louder) for the same power input. Thus, a 100 W driver ("A") rated at 92 dB for 1 W @ 1 m sensitivity puts out twice as much acoustic power as a 200 W driver ("B") rated at 89 dB for 1 W @ 1 m when both are driven with 100 W of input power. In this particular example, when driven at 100 W, speaker A produces the same SPL, or loudness as speaker B would produce with 200 W input. Thus, a 3 dB increase in sensitivity of the speaker means that it needs half the amplifier power to achieve a given SPL. This translates into a smaller, less complex power amplifier—and often, to reduced overall system cost.

It is typically not possible to combine high efficiency (especially at low frequencies) with compact enclosure size and adequate low frequency response. One can, for the most part, choose only two of the three parameters when designing a speaker system. So, for example, if extended low-frequency performance and small box size are important, one must accept low efficiency. This rule of thumb is sometimes called Hofmann's Iron Law (after J.A. Hofmann, the "H" in KLH).

Listening Environment

The interaction of a loudspeaker system with its environment is complex and is largely out of the loudspeaker designer's control. Most listening rooms present a more or less reflective environment, depending on size, shape, volume, and furnishings. This means the sound reaching a listener's ears consists not only of sound directly from the speaker system, but also the same sound delayed by traveling to and from (and being modified by) one or more surfaces. These reflected sound waves, when added to the direct sound, cause cancellation and addition at assorted frequencies (e.g., from resonant room modes), thus changing the timbre and character of the sound at the listener's ears. The human brain is very sensitive to small variations, including some of these, and this is part of the reason why a loudspeaker system sounds different at different listening positions or in different rooms.

A significant factor in the sound of a loudspeaker system is the amount of absorption and diffusion present in the environment. Clapping one's hands in a typical empty room, without draperies or carpet, produces a zippy, fluttery echo due both to a lack of absorption and to reverberation (that is, repeated echoes) from flat reflective walls, floor, and ceiling. The addition of hard surfaced furniture, wall hangings, shelving and even baroque plaster ceiling decoration changes the echoes, primarily because of diffusion caused by reflective objects with shapes and surfaces having sizes on the order of the sound wavelengths. This somewhat breaks up the simple reflections otherwise caused by bare flat surfaces, and spreads the reflected energy of an incident wave over a larger angle on reflection.

Placement

In a typical rectangular listening room, the hard, parallel surfaces of the walls, floor and ceiling cause primary acoustic resonance nodes in each of the three dimensions: left-right, up-down and forward-backward. Furthermore, there are more complex resonance modes involving three, four, five and even all six boundary surfaces combining to create standing waves. Low frequencies excite these modes the most, since long wavelengths are not much affected by furniture compositions or placement. The mode spacing is critical, especially in small and medium size rooms like recording studios, home theaters and broadcast studios. The proximity of the loudspeakers to room boundaries affects how strongly the resonances are excited as well as affecting the relative strength at each frequency. The location of the listener is critical, too, as a position near a boundary can have a great effect on the perceived balance of frequencies. This is because standing wave patterns are most easily heard in these locations and at lower frequencies, below the Schroeder frequency – typically around 200–300 Hz, depending on room size.

Directivity

Acousticians, in studying the radiation of sound sources have developed some concepts important to understanding how loudspeakers are perceived. The simplest possible radiating source is a point source, sometimes called a simple source. An ideal point source is an infinitesimally small point radiating sound. It may be easier to imagine a tiny pulsating sphere, uniformly increasing and decreasing in diameter, sending out sound waves in all directions equally, independent of frequency.

Any object radiating sound, including a loudspeaker system, can be thought of as being composed of combinations of such simple point sources. The radiation pattern of a combination of point sources is not the same as for a single source, but depends on the distance and orientation between the sources, the position relative to them from which the listener hears the combination, and the frequency of the sound involved. Using geometry and calculus, some simple combinations of sources are easily solved; others are not.

One simple combination is two simple sources separated by a distance and vibrating out of phase, one miniature sphere expanding while the other is contracting. The pair is known as a doublet, or dipole, and the radiation of this combination is similar to that of a very small dynamic loudspeaker operating without a baffle. The directivity of a dipole is a figure 8 shape with maximum output along a vector that connects the two sources and minimums to the sides when the observing point is equidistant from the two sources, where the sum of the positive and negative waves cancel each other. While most drivers are dipoles, depending on the enclosure to which they are attached, they may radiate as monopoles, dipoles (or bipoles). If mounted on a finite baffle, and these out of phase waves are allowed to interact, dipole peaks and nulls in the frequency response result. When the rear radiation is absorbed or trapped in a box, the diaphragm becomes a monopole radiator. Bipolar speakers, made by mounting in-phase monopoles (both moving out of or into the box in unison) on opposite sides of a box, are a method of approaching omnidirectional radiation patterns.

In real life, individual drivers are complex 3D shapes such as cones and domes, and they are placed on a baffle for various reasons. A mathematical expression for the directivity of a complex shape, based on modeling combinations of point sources, is usually not possible, but in the far field, the directivity of a loudspeaker with a circular diaphragm is close to that of a flat circular piston, so it can be used as an illustrative simplification for discussion. As a simple example of the mathematical physics involved, consider the following: the formula for far field directivity of a flat circular piston in an infinite baffle is $p(\theta) = \dfrac{p_0 J_1(k_a \sin \theta)}{k_a \sin \theta}$ where $k_a = \dfrac{2\pi a}{\lambda}$, p_0 is the pressure on axis, a is the piston radius, λ is the wavelength (i.e. $\lambda = \dfrac{c}{f} = \dfrac{\text{speed of sound}}{\text{frequency}}$) θ is the angle off axis and J_1 is the Bessel function of the first kind.

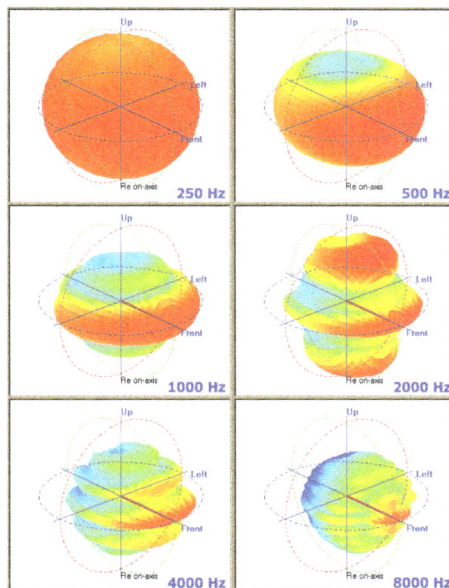

Polar plots of a four-driver industrial columnar public address loudspeaker taken at six frequencies. Note how the pattern is nearly omnidirectional at low frequencies, converging to a wide fan-shaped pattern at 1 kHz, then separating into lobes and getting weaker at higher frequencies

A planar source radiates sound uniformly for low frequencies' wavelengths longer than the dimensions of the planar source, and as frequency increases, the sound from such a source focuses into an increasingly narrower angle. The smaller the driver, the higher the frequency where this narrowing of directivity occurs. Even if the diaphragm is not perfectly circular, this effect occurs such that larger sources are more directive. Several loudspeaker designs approximate this behavior. Most are electrostatic or planar magnetic designs.

Various manufacturers use different driver mounting arrangements to create a specific type of sound field in the space for which they are designed. The resulting radiation patterns may be intended to more closely simulate the way sound is produced by real instruments, or simply create a controlled energy distribution from the input signal (some using this approach are called monitors, as they are useful in checking the signal just recorded in a studio). An example of the first is a room corner system with many small drivers on the surface of a 1/8 sphere. A system design of this type was patented and produced commercially by Professor Amar Bose—the 2201. Later Bose models have deliberately emphasized production of both direct and reflected sound by the loudspeaker itself, regardless of its environment. The designs are controversial in high fidelity circles, but have proven commercially successful. Several other manufacturers' designs follow similar principles.

Directivity is an important issue because it affects the frequency balance of sound a listener hears, and also the interaction of the speaker system with the room and its contents. A very directed speaker (i.e., on an axis perpendicular to the speaker face) may result in a reverberant field lacking in high frequencies, giving the impression the speaker is deficient in treble even though it measures well on axis (e.g., "flat" across the entire frequency range). Speakers with very wide, or rapidly increasing directivity at high frequencies, can give the impression that there is too much treble (if the listener is on axis) or too little (if the listener is off axis). This is part of the reason why on-axis frequency response measurement is not a complete characterization of the sound of a given loudspeaker.

Other Speaker Designs

While dynamic cone speakers remain the most popular choice, many other speaker technologies exist.

With a Diaphragm

Moving-iron Loudspeakers

Unlike the newer dynamic (moving coil) design, a moving-iron speaker uses a stationary coil to vibrate a magnetized piece of metal (called the iron, reed, or armature). The metal is either attached to the diaphragm, or is the diaphragm itself. This design was the original loudspeaker design, dating back to the early telephone. Moving coil drivers are inefficient and can only produce a small band of sound. They require large magnets and coils to increase force.

Balanced armature drivers (a type of moving iron driver) use an armature that moves like a see-saw or diving board. Since they are not damped, they are highly efficient, but they also produce strong resonances. They are still used today for high end earphones and hearing aids, where small size and high efficiency are important.

Piezoelectric Speakers

Piezoelectric speakers are frequently used as beepers in watches and other electronic devices, and are sometimes used as tweeters in less-expensive speaker systems, such as computer speakers and portable radios. Piezoelectric speakers have several advantages over conventional loudspeakers: they are resistant to overloads that would normally destroy most high frequency drivers, and they can be used without a crossover due to their electrical properties. There are also disadvantages: some amplifiers can oscillate when driving capacitive loads like most piezoelectrics, which results in distortion or damage to the amplifier. Additionally, their frequency response, in most cases, is inferior to that of other technologies. This is why they are generally used in single frequency (beeper) or non-critical applications.

Piezoelectric speakers can have extended high frequency output, and this is useful in some specialized circumstances; for instance, sonar applications in which piezoelectric variants are used as both output devices (generating underwater sound) and as input devices (acting as the sensing components of underwater microphones). They have advantages in these applications, not the least of which is simple and solid state construction that resists seawater better than a ribbon or cone based device would.

In 2013, Kyocera introduced piezoelectric ultra-thin medium-size film speakers with only 1 milimeter of thickness and 7 grams of weight for their 55" OLED televisions and they hope the speakers will also be used in PCs and tablets. Besides medium-size, there are also large and small sizes which can all produce relatively the same quality of sound and volume within 180 degrees. The highly responsive speaker material provides better clarity than traditional TV speakers.

Magnetostatic Loudspeakers

Instead of a voice coil driving a speaker cone, a magnetostatic speaker uses an array of metal strips bonded to a large film membrane. The magnetic field produced by signal current flowing through the strips interacts with the field of permanent bar magnets mounted behind them. The force produced moves the membrane and so the air in front of it. Typically, these designs are less efficient than conventional moving-coil speakers.

Magnetostrictive Speakers

Magnetostrictive transducers, based on magnetostriction, have been predominantly used as sonar ultrasonic sound wave radiators, but their use has spread also to audio speaker systems. Magnetostrictive speaker drivers have some special advantages: they can provide greater force (with smaller excursions) than other technologies; low excursion can avoid distortions from large excursion as in other designs; the magnetizing coil is stationary and therefore more easily cooled; they are robust because delicate suspensions and voice coils are not required. Magnetostrictive speaker modules have been produced by Fostex and FeONIC and subwoofer drivers have also been produced.

Electrostatic Loudspeakers

Electrostatic loudspeakers use a high voltage electric field (rather than a magnetic field) to drive a thin statically charged membrane. Because they are driven over the entire membrane surface rather than from a small voice coil, they ordinarily provide a more linear and lower-distortion motion than dynamic drivers. They also have a relatively narrow dispersion pattern that can make for precise sound-field positioning. However, their optimum listening area is small and they are not very efficient speakers. They have the disadvantage that the diaphragm excursion is severely limited because of practical construction limitations—the further apart the stators are positioned, the higher the voltage must be to achieve acceptable efficiency. This increases the tendency for electrical arcs as well as increasing the speaker's attraction of dust particles. Arcing remains a potential problem with current technologies, especially when the panels are allowed to collect dust or dirt and are driven with high signal levels.

Electrostatics are inherently dipole radiators and due to the thin flexible membrane are less suited for use in enclosures to reduce low frequency cancellation as with common cone drivers. Due to this and the low excursion capability, full range electrostatic loudspeakers are large by nature, and the bass rolls off at a frequency corresponding to a quarter wavelength of the narrowest panel dimension. To reduce the size of commercial products, they are sometimes used as a high frequency driver in combination with a conventional dynamic driver that handles the bass frequencies effectively.

Electrostatics are usually driven through a step-up transformer that multiplies the voltage swings produced by the power amplifier. This transformer also multiplies the capacitive load that is inherent in electrostatic transducers, which means the effective impedance presented to the power amplifiers varies widely by frequency. A speaker that is nominally 8 ohms may actually present a load of 1 ohm at higher frequencies, which is challenging to some amplifier designs.

Ribbon and Planar Magnetic Loudspeakers

A ribbon speaker consists of a thin metal-film ribbon suspended in a magnetic field. The electrical signal is applied to the ribbon, which moves with it to create the sound. The advantage of a ribbon driver is that the ribbon has very little mass; thus, it can accelerate very quickly, yielding very good high-frequency response. Ribbon loudspeakers are often very fragile—some can be torn by a strong gust of air. Most ribbon tweeters emit sound in a dipole pattern. A few have backings that limit the dipole radiation pattern. Above and below the ends of the more or less rectangular ribbon, there is less audible output due to phase cancellation, but the precise amount of directivity depends on ribbon length. Ribbon designs generally require exceptionally powerful magnets, which makes them costly to manufacture. Ribbons have a very low resistance that most amplifiers cannot drive directly. As a result, a step down transformer is typically used to increase the current through the ribbon. The amplifier "sees" a load that is the ribbon's resistance times the transformer turns ratio squared. The transformer must be carefully designed so that its frequency response and parasitic losses do not degrade the sound, further increasing cost and complication relative to conventional designs.

Planar magnetic speakers (having printed or embedded conductors on a flat diaphragm) are sometimes described as ribbons, but are not truly ribbon speakers. The term planar is generally reserved for speakers with roughly rectangular flat surfaces that radiate in a bipolar (i.e., front and back) manner. Planar magnetic speakers consist of a flexible membrane with a voice coil printed or mounted on it. The current flowing through the coil interacts with the magnetic field of carefully placed magnets on either side of the diaphragm, causing the membrane to vibrate more or less uniformly and without much bending or wrinkling. The driving force covers a large percentage of the membrane surface and reduces resonance problems inherent in coil-driven flat diaphragms.

Bending Wave Loudspeakers

Bending wave transducers use a diaphragm that is intentionally flexible. The rigidity of the material increases from the center to the outside. Short wavelengths radiate primarily from the inner area, while longer waves reach the edge of the speaker. To prevent reflections from the outside back into the center, long waves are absorbed by a surrounding damper. Such transducers can cover a wide frequency range (80 Hz to 35,000 Hz) and have been promoted as being close to an ideal point sound source. This uncommon approach is being taken by only a very few manufacturers, in very different arrangements.

The Ohm Walsh loudspeakers use a unique driver designed by Lincoln Walsh, who had been a radar development engineer in WWII. He became interested in audio equipment design and his last project was a unique, one-way speaker using a single driver. The cone faced down into a sealed, airtight enclosure. Rather than move back-and-forth as conventional speakers do, the cone rippled and created sound in a manner known in RF electronics as a "transmission line". The new speaker created a cylindrical sound field. Lincoln Walsh died before his speaker was released to the public. The Ohm Acoustics firm has produced several loudspeaker models using the Walsh driver design since then.

The German firm, Manger, has designed and produced a bending wave driver that at first glance appears conventional. In fact, the round panel attached to the voice coil bends in a carefully controlled way to produce full range sound. Josef W. Manger was awarded with the "Diesel Medal" for extraordinary developments and inventions by the German institute of inventions.

Flat Panel Loudspeakers

There have been many attempts to reduce the size of speaker systems, or alternatively to make them less obvious. One such attempt was the development of "exciter" transducer coils mounted to flat panels to act as sound sources, most accurately called exciter/panel drivers. These can then be made in a neutral color and hung on walls where they are less noticeable than many speakers, or can be deliberately painted with patterns, in which case they can function decoratively. There are two related problems with flat panel techniques: first, a flat panel is necessarily more flexible than a cone shape in the same material, and therefore moves as a single unit even less, and second, resonances in the panel are difficult to control, leading to considerable distortions. Some progress has been made using such lightweight, rigid, materials such as Styrofoam, and there have been several flat panel systems commercially produced in recent years.

Heil Air Motion Transducers

Oskar Heil invented the air motion transducer in the 1960s. In this approach, a pleated diaphragm is mounted in a magnetic field and forced to close and open under control of a music signal. Air is forced from between the pleats in accordance with the imposed signal, generating sound. The drivers are less fragile than ribbons and considerably more efficient (and able to produce higher absolute output levels) than ribbon, electrostatic, or planar magnetic tweeter designs.

ESS, a California manufacturer, licensed the design, employed Heil, and produced a range of speaker systems using his tweeters during the 1970s and 1980s. Lafayette Radio, a large US retail store chain, also sold speaker systems using such tweeters for a time. There are several manufacturers of these drivers (at least two in Germany—one of which produces a range of high-end professional speakers using tweeters and mid-range drivers based on the technology) and the drivers are increasingly used in professional audio. Martin Logan produces several AMT speakers in the US. GoldenEar Technologies incorporates them in its entire speaker line.

Transparent Ionic Conduction Speaker

In 2013, a research team introduced Transparent ionic conduction speaker which a 2 layers transparent conductive gel and a layer of transparent rubber in between to make high voltage and high actuation work to reproduce good sound quality. The speaker is suitable for robotics, mobile computing and adaptive optics fields.

Without A Diaphragm

Plasma Arc Speakers

Plasma arc loudspeakers use electrical plasma as a radiating element. Since plasma has minimal mass, but is charged and therefore can be manipulated by an electric field, the result is a very linear output at frequencies far higher than the audible range. Problems of maintenance and reliability for this approach tend to make it unsuitable for mass market use. In 1978 Alan E. Hill of the Air Force Weapons Laboratory in Albuquerque, NM, designed the Plasmatronics Hill Type I, a tweeter whose plasma was generated from helium gas. This avoided the ozone and nitrous oxide produced by RF decomposition of air in an earlier generation of plasma tweeters made by the pioneering DuKane Corporation, who produced the Ionovac (marketed as the Ionofane in the UK) during the 1950s. Currently, there remain a few manufacturers in Germany who use this design, and a do-it-yourself design has been published and has been available on the Internet.

A less expensive variation on this theme is the use of a flame for the driver, as flames contain ionized (electrically charged) gases.

Thermoacoustic Speakers

In 2008, researchers of Tsinghua University demonstrated a thermoacoustic loudspeaker of carbon nanotube thin film, whose working mechanism is a thermoacoustic effect. Sound frequency electric currents are used to periodically heat the CNT and thus result in sound generation in the surrounding air. The CNT thin film loudspeaker is transparent, stretchable and flexible. In 2013, researchers of Tsinghua University further present a thermoacoustic earphone of carbon nanotube thin yarn and a thermoacoustic surface-mounted device. They are both fully integrated devices and compatible with Si-based semiconducting technology.

Rotary Woofers

A rotary woofer is essentially a fan with blades that constantly change their pitch, allowing them to easily push the air back and forth. Rotary woofers are able to efficiently reproduce infrasound frequencies, which are difficult to impossible to achieve on a traditional speaker with a diaphragm. They are often employed in movie theaters to recreate rumbling bass effects, such as explosions.

New Technologies for Driving Speakers

Digital Speakers

Digital speakers have been the subject of experiments performed by Bell Labs as far back as the 1920s. The design is simple; each bit controls a driver, which is either fully 'on' or 'off'. Problems with this design have led manufacturers to abandon it as impractical for the present. First, for a reasonable number of bits (required for adequate sound reproduction quality), the physical size of a speaker system becomes very large. Secondly, due to inherent analog digital conversion problems, the effect of aliasing is unavoidable, so that the audio output is "reflected" at equal amplitude in the frequency domain, on the other side of the sampling frequency, causing an unacceptably high level of ultrasonics to accompany the desired output. No workable scheme has been found to adequately deal with this.

The term "digital" or "digital-ready" is often used for marketing purposes on speakers or headphones, but these systems are not digital in the sense described above. Rather, they are conventional speakers that can be used with digital sound sources (e.g., optical media, MP3 players, etc.), as can any conventional speaker.

Microphone

A microphone, colloquially nicknamed mic or mike is a transducer that converts sound into an electrical signal.

An AKG C214 condenser microphone with shock mount

Microphones are used in many applications such as telephones, hearing aids, public address systems for concert halls and public events, motion picture production, live and recorded audio engineering, two-way radios, megaphones, radio and television broadcasting, and in computers for recording voice, speech recognition, VoIP, and for non-acoustic purposes such as ultrasonic checking or knock sensors.

A Sennheiser dynamic microphone

Most microphones today use electromagnetic induction (dynamic microphones), capacitance change (condenser microphones) or piezoelectricity (piezoelectric microphones) to produce an electrical signal from air pressure variations. Microphones typically need to be connected to a pre-amplifier before the signal can be recorded or reproduced.

History

In order to speak to larger groups of people, there was a desire to increase the volume of the spoken word. The earliest known device to achieve this dates to 600 BC with the invention of masks with specially designed mouth openings that acoustically augmented the voice in amphitheatres. In 1665, the English physicist Robert Hooke was the first to experiment with a medium other than air with the invention of the "lovers' telephone" made of stretched wire with a cup attached at each end.

German inventor Johann Philipp Reis designed an early sound transmitter that used a metallic strip attached to a vibrating membrane that would produce intermittent current. Better results were achieved with the "liquid transmitter" design in Scottish-American Alexander Graham Bell's telephone of 1876 – the diaphragm was attached to a conductive rod in an acid solution. These systems, however, gave a very poor sound quality.

David Edward Hughes invented a carbon microphone in the 1870s.

The first microphone that enabled proper voice telephony was the (loose-contact) carbon microphone. This was independently developed by David Edward Hughes in England and Emile Berliner and Thomas Edison in the US. Although Edison was awarded the first patent (after a long legal dispute) in mid-1877, Hughes had demonstrated his working device in front of many witnesses some years earlier, and most historians credit him with its invention. The carbon microphone is the direct prototype of today's microphones and was critical in the development of telephony, broadcasting and the recording industries. Thomas Edison refined the carbon microphone into his carbon-button transmitter of 1886. This microphone was employed at the first ever radio broadcast, a performance at the New York Metropolitan Opera House in 1910.

Jack Brown interviews Humphrey Bogart and Lauren Bacall for broadcast to troops overseas during World War II.

In 1916, C. Wente of Bell Labs developed the next breakthrough with the first condenser microphone. In 1923, the first practical moving coil microphone was built. "The Marconi Skykes" or "magnetophon", developed by Captain H. J. Round, was the standard for BBC studios in London. This was improved in 1930 by Alan Blumlein and Herbert Holman who released the HB1A and was the best standard of the day.

Also in 1923, the ribbon microphone was introduced, another electromagnetic type, believed to have been developed by Harry F. Olson, who essentially reverse-engineered a ribbon speaker. Over the years these microphones were developed by several companies, most notably RCA that made large advancements in pattern control, to give the microphone directionality. With television and film technology booming there was demand for high fidelity microphones and greater directionality. Electro-Voice responded with their Academy Award-winning shotgun microphone in 1963.

During the second half of 20th century development advanced quickly with the Shure Brothers bringing out the SM58 and SM57. Digital was pioneered by Milab in 1999 with the DM-1001. The latest research developments include the use of fibre optics, lasers and interferometers.

Components

Electronic symbol for a microphone

The sensitive transducer element of a microphone is called its *element* or *capsule*. Except in thermophone based microphones, sound is first converted to mechanical motion by means of a diaphragm, the motion of which is then converted to an electrical signal. A complete microphone also includes a housing, some means of bringing the signal from the element to other equipment, and often an electronic circuit to adapt the output of the capsule to the equipment being driven. A wireless microphone contains a radio transmitter.

Varieties

Microphones are referred to by their transducer principle, such as condenser, dynamic, etc., and by their directional characteristics. Sometimes other characteristics such as diaphragm size, intended use or orientation of the principal sound input to the principal axis (end- or side-address) of the microphone are used to describe the microphone.

Condenser Microphone

The condenser microphone, invented at Bell Labs in 1916 by E. C. Wente, is also called a capacitor microphone or electrostatic microphone—capacitors were historically called condensers. Here, the diaphragm acts as one plate of a capacitor, and the vibrations produce changes in the distance between the plates. There are two types, depending on the method of extracting the audio signal from the transducer: DC-biased microphones, and radio frequency (RF) or high frequency (HF) condenser microphones. With a DC-biased microphone, the plates are biased with a fixed charge (Q). The voltage maintained across the capacitor plates changes with the vibrations in the air, according to the capacitance equation ($C = Q/V$), where Q = charge in coulombs, C = capacitance in farads and V = potential difference in volts. The capacitance of the plates is inversely proportional to the distance between them for a parallel-plate capacitor. The assembly of fixed and movable plates is called an "element" or "capsule".

Inside the Oktava 319 condenser microphone

A nearly constant charge is maintained on the capacitor. As the capacitance changes, the charge across the capacitor does change very slightly, but at audible frequencies it is sensibly constant. The capacitance of the capsule (around 5 to 100 pF) and the value of the bias resistor (100 MΩ to tens of GΩ) form a filter that is high-pass for the audio signal, and low-pass for the bias voltage. Note that the time constant of an RC circuit equals the product of the resistance and capacitance.

Within the time-frame of the capacitance change (as much as 50 ms at 20 Hz audio signal), the charge is practically constant and the voltage across the capacitor changes instantaneously to reflect the change in capacitance. The voltage across the capacitor varies above and below the bias voltage. The voltage difference between the bias and the capacitor is seen across the series resistor. The voltage across the resistor is amplified for performance or recording. In most cases, the electronics in the microphone itself contribute no voltage gain as the voltage differential is quite significant, up to several volts for high sound levels. Since this is a very high impedance circuit, current gain only is usually needed, with the voltage remaining constant.

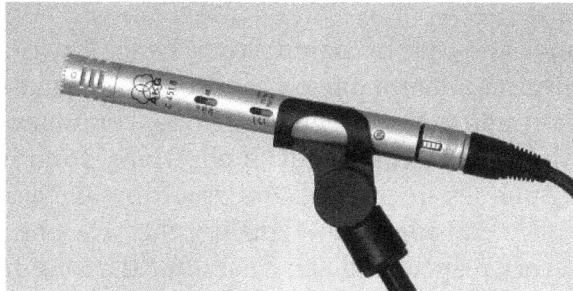

AKG C451B small-diaphragm condenser microphone

RF condenser microphones use a comparatively low RF voltage, generated by a low-noise oscillator. The signal from the oscillator may either be amplitude modulated by the capacitance changes produced by the sound waves moving the capsule diaphragm, or the capsule may be part of a resonant circuit that modulates the frequency of the oscillator signal. Demodulation yields a low-noise audio frequency signal with a very low source impedance. The absence of a high bias voltage permits the use of a diaphragm with looser tension, which may be used to achieve wider frequency response due to higher compliance. The RF biasing process results in a lower electrical impedance capsule, a useful by-product of which is that RF condenser microphones can be operated in damp weather conditions that could create problems in DC-biased microphones with contaminated insulating surfaces. The Sennheiser "MKH" series of microphones use the RF biasing technique.

Condenser microphones span the range from telephone transmitters through inexpensive karaoke microphones to high-fidelity recording microphones. They generally produce a high-quality audio signal and are now the popular choice in laboratory and recording studio applications. The inherent suitability of this technology is due to the very small mass that must be moved by the incident sound wave, unlike other microphone types that require the sound wave to do more work. They require a power source, provided either via microphone inputs on equipment as phantom power or from a small battery. Power is necessary for establishing the capacitor plate voltage, and is also needed to power the microphone electronics (impedance conversion in the case of electret and DC-polarized microphones, demodulation or detection in the case of RF/HF microphones). Condenser microphones are also available with two diaphragms that can be electrically connected to provide a range of polar patterns, such as cardioid, omnidirectional, and figure-eight. It is also possible to vary the pattern continuously with some microphones, for example the Røde NT2000 or CAD M179.

A valve microphone is a condenser microphone that uses a vacuum tube (valve) amplifier. They remain popular with enthusiasts of tube sound.

Electret Condenser Microphone

An electret microphone is a type of capacitor microphone invented by Gerhard Sessler and Jim West at Bell laboratories in 1962. The externally applied charge described above under condenser microphones is replaced by a permanent charge in an electret material. An electret is a ferroelectric material that has been permanently electrically charged or *polarized*. The name comes from *electro*static and magn*et*; a static charge is embedded in an electret by alignment of the static charges in the material, much the way a magnet is made by aligning the magnetic domains in a piece of iron.

Due to their good performance and ease of manufacture, hence low cost, the vast majority of microphones made today are electret microphones; a semiconductor manufacturer estimates annual production at over one billion units. Nearly all cell-phone, computer, PDA and headset microphones are electret types. They are used in many applications, from high-quality recording and lavalier use to built-in microphones in small sound recording devices and telephones. Though electret microphones were once considered low quality, the best ones can now rival traditional condenser microphones in every respect and can even offer the long-term stability and ultra-flat response needed for a measurement microphone. Unlike other capacitor microphones, they require no polarizing voltage, but often contain an integrated preamplifier that does require power (often incorrectly called polarizing power or bias). This preamplifier is frequently phantom powered in sound reinforcement and studio applications. Monophonic microphones designed for personal computer (PC) use, sometimes called multimedia microphones, use a 3.5 mm plug as usually used, without power, for stereo; the ring, instead of carrying the signal for a second channel, carries power via a resistor from (normally) a 5 V supply in the computer. Stereophonic microphones use the same connector; there is no obvious way to determine which standard is used by equipment and microphones.

Only the best electret microphones rival good DC-polarized units in terms of noise level and quality; electret microphones lend themselves to inexpensive mass-production, while inherently expensive non-electret condenser microphones are made to higher quality.

Dynamic Microphone

Dynamic microphones (also known as moving-coil microphones) work via electromagnetic induction. They are robust, relatively inexpensive and resistant to moisture. This, coupled with their potentially high gain before feedback, makes them ideal for on-stage use.

Patti Smith singing into a Shure SM58 (dynamic cardioid type) microphone

Dynamic microphones use the same dynamic principle as in a loudspeaker, only reversed. A small movable induction coil, positioned in the magnetic field of a permanent magnet, is attached to the diaphragm. When sound enters through the windscreen of the microphone, the sound wave moves the diaphragm. When the diaphragm vibrates, the coil moves in the magnetic field, producing a varying current in the coil through electromagnetic induction. A single dynamic membrane does not respond linearly to all audio frequencies. Some microphones for this reason utilize multiple membranes for the different parts of the audio spectrum and then combine the resulting signals. Combining the multiple signals correctly is difficult and designs that do this are rare and tend to be expensive. There are on the other hand several designs that are more specifically aimed towards isolated parts of the audio spectrum. The AKG D 112, for example, is designed for bass response rather than treble. In audio engineering several kinds of microphones are often used at the same time to get the best results.

Ribbon Microphone

Ribbon microphones use a thin, usually corrugated metal ribbon suspended in a magnetic field. The ribbon is electrically connected to the microphone's output, and its vibration within the magnetic field generates the electrical signal. Ribbon microphones are similar to moving coil microphones in the sense that both produce sound by means of magnetic induction. Basic ribbon microphones detect sound in a bi-directional (also called figure-eight, as in the diagram below) pattern because the ribbon is open on both sides. Also, because the ribbon is much less mass it responds to the air velocity rather than the sound pressure. Though the symmetrical front and rear pickup can be a nuisance in normal stereo recording, the high side rejection can be used to advantage by positioning a ribbon microphone horizontally, for example above cymbals, so that the rear lobe picks up only sound from the cymbals. Crossed figure 8, or Blumlein pair, stereo recording is gaining in popularity, and the figure-eight response of a ribbon microphone is ideal for that application.

Edmund Lowe using a ribbon microphone

Other directional patterns are produced by enclosing one side of the ribbon in an acoustic trap or baffle, allowing sound to reach only one side. The classic RCA Type 77-DX microphone has several externally adjustable positions of the internal baffle, allowing the selection of several response patterns ranging from "figure-eight" to "unidirectional". Such older ribbon microphones, some of which still provide high quality sound reproduction, were once valued for this reason, but a good low-frequency response could only be obtained when the ribbon was suspended very loosely, which made them relatively fragile. Modern ribbon materials, including new nanomaterials have now been introduced that eliminate those concerns, and even improve the effective dynamic range of ribbon microphones at low frequencies. Protective wind screens can reduce the danger of damaging a vintage ribbon, and also reduce plosive artifacts in the recording. Properly designed wind screens produce negligible treble attenuation. In common with other classes of dynamic microphone, ribbon microphones don't require phantom power; in fact, this voltage can damage some older ribbon microphones. Some new modern ribbon microphone designs incorporate a preamplifier and, therefore, do require phantom power, and circuits of modern passive ribbon microphones, *i.e.*, those without the aforementioned preamplifier, are specifically designed to resist damage to the ribbon and transformer by phantom power. Also there are new ribbon materials available that are immune to wind blasts and phantom power.

Carbon Microphone

A carbon microphone, also known as a carbon button microphone (or sometimes just a button microphone), uses a capsule or button containing carbon granules pressed between two metal plates like the Berliner and Edison microphones. A voltage is applied across the metal plates, causing a small current to flow through the carbon. One of the plates, the diaphragm, vibrates in sympathy with incident sound waves, applying a varying pressure to the carbon. The changing pressure deforms the granules, causing the contact area between each pair of adjacent granules to change, and this causes the electrical resistance of the mass of granules to change. The changes in resistance cause a corresponding change in the current flowing through the microphone, producing the electrical signal. Carbon microphones were once commonly used in telephones; they have extremely low-quality sound reproduction and a very limited frequency response range, but are very robust devices. The Boudet microphone, which used relatively large carbon balls, was similar to the granule carbon button microphones.

Unlike other microphone types, the carbon microphone can also be used as a type of amplifier, using a small amount of sound energy to control a larger amount of electrical energy. Carbon microphones found use as early telephone repeaters, making long distance phone calls possible in the era before vacuum tubes. These repeaters worked by mechanically coupling a magnetic telephone receiver to a carbon microphone: the faint signal from the receiver was transferred to the microphone, where it modulated a stronger electric current, producing a stronger electrical signal to send down the line. One illustration of this amplifier effect was the oscillation caused by feedback, resulting in an audible squeal from the old "candlestick" telephone if its earphone was placed near the carbon microphone.

Piezoelectric Microphone

A crystal microphone or piezo microphone uses the phenomenon of piezoelectricity—the ability of some materials to produce a voltage when subjected to pressure—to convert vibrations into an electrical signal. An example of this is potassium sodium tartrate, which is a piezoelectric crystal that works as a transducer, both as a microphone and as a slimline loudspeaker component. Crystal microphones were once commonly supplied with vacuum tube (valve) equipment, such as domestic tape recorders. Their high output impedance matched the high input impedance (typically about 10 megohms) of the vacuum tube input stage well. They were difficult to match to early transistor equipment, and were quickly supplanted by dynamic microphones for a time, and later small electret condenser devices. The high impedance of the crystal microphone made it very susceptible to handling noise, both from the microphone itself and from the connecting cable.

Piezoelectric transducers are often used as contact microphones to amplify sound from acoustic musical instruments, to sense drum hits, for triggering electronic samples, and to record sound in challenging environments, such as underwater under high pressure. Saddle-mounted pickups on acoustic guitars are generally piezoelectric devices that contact the strings passing over the saddle. This type of microphone is different from magnetic coil pickups commonly visible on typical electric guitars, which use magnetic induction, rather than mechanical coupling, to pick up vibration.

Fiber Optic Microphone

A fiber optic microphone converts acoustic waves into electrical signals by sensing changes in light intensity, instead of sensing changes in capacitance or magnetic fields as with conventional microphones.

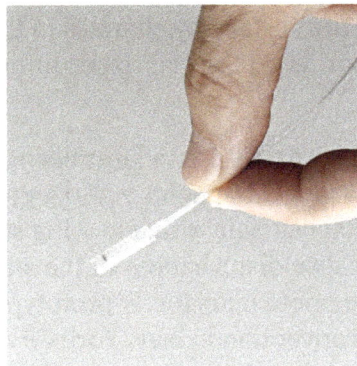

The Optoacoustics 1140 fiber optic microphone

During operation, light from a laser source travels through an optical fiber to illuminate the surface of a reflective diaphragm. Sound vibrations of the diaphragm modulate the intensity of light reflecting off the diaphragm in a specific direction. The modulated light is then transmitted over a second optical fiber to a photo detector, which transforms the intensity-modulated light into analog or digital audio for transmission or recording. Fiber optic microphones possess high dynamic and frequency range, similar to the best high fidelity conventional microphones.

Fiber optic microphones do not react to or influence any electrical, magnetic, electrostatic or radioactive fields (this is called EMI/RFI immunity). The fiber optic microphone design is therefore ideal for use in areas where conventional microphones are ineffective or dangerous, such as inside industrial turbines or in magnetic resonance imaging (MRI) equipment environments.

Fiber optic microphones are robust, resistant to environmental changes in heat and moisture, and can be produced for any directionality or impedance matching. The distance between the microphone's light source and its photo detector may be up to several kilometers without need for any preamplifier or other electrical device, making fiber optic microphones suitable for industrial and surveillance acoustic monitoring.

Fiber optic microphones are used in very specific application areas such as for infrasound monitoring and noise-canceling. They have proven especially useful in medical applications, such as allowing radiologists, staff and patients within the powerful and noisy magnetic field to converse normally, inside the MRI suites as well as in remote control rooms. Other uses include industrial equipment monitoring and audio calibration and measurement, high-fidelity recording and law enforcement.

Laser Microphone

Laser microphones are often portrayed in movies as spy gadgets, because they can be used to pick up sound at a distance from the microphone equipment. A laser beam is aimed at the surface of a window or other plane surface that is affected by sound. The vibrations of this surface change the angle at which the beam is reflected, and the motion of the laser spot from the returning beam is detected and converted to an audio signal.

In a more robust and expensive implementation, the returned light is split and fed to an interferometer, which detects movement of the surface by changes in the optical path length of the reflected beam. The former implementation is a tabletop experiment; the latter requires an extremely stable laser and precise optics.

A new type of laser microphone is a device that uses a laser beam and smoke or vapor to detect sound vibrations in free air. On 25 August 2009, U.S. patent 7,580,533 issued for a Particulate Flow Detection Microphone based on a laser-photocell pair with a moving stream of smoke or vapor in the laser beam's path. Sound pressure waves cause disturbances in the smoke that in turn cause variations in the amount of laser light reaching the photo detector. A prototype of the device was demonstrated at the 127th Audio Engineering Society convention in New York City from 9 through 12 October 2009.

Liquid Microphone

Early microphones did not produce intelligible speech, until Alexander Graham Bell made improvements including a variable-resistance microphone/transmitter. Bell's liquid transmitter consisted of a metal cup filled with water with a small amount of sulfuric acid added. A sound wave caused the diaphragm to move, forcing a needle to move up and down in the water. The electrical resistance between the wire and the cup was then inversely proportional to the size of the water meniscus around the submerged needle. Elisha Gray filed a caveat for a version using a brass rod instead of the needle. Other minor variations and improvements were made to the liquid microphone by Majoranna, Chambers, Vanni, Sykes, and Elisha Gray, and one version was patented by Reginald Fessenden in 1903. These were the first working microphones, but they were not practical for commercial application. The famous first phone conversation between Bell and Watson took place using a liquid microphone.

MEMS Microphone

The MEMS (MicroElectrical-Mechanical System) microphone is also called a microphone chip or silicon microphone. A pressure-sensitive diaphragm is etched directly into a silicon wafer by

MEMS processing techniques, and is usually accompanied with integrated preamplifier. Most MEMS microphones are variants of the condenser microphone design. Digital MEMS microphones have built in analog-to-digital converter (ADC) circuits on the same CMOS chip making the chip a digital microphone and so more readily integrated with modern digital products. Major manufacturers producing MEMS silicon microphones are Wolfson Microelectronics (WM7xxx) now Cirrus Logic, InvenSense (product line sold by Analog Devices), Akustica (AKU200x), Infineon (SMM310 product), Knowles Electronics, Memstech (MSMx), NXP Semiconductors (division bought by Knowles), Sonion MEMS, Vesper, AAC Acoustic Technologies, and Omron.

More recently, there has been increased interest and research into making piezoelectric MEMS microphones which are a significant architectural and material change from existing condenser style MEMS designs.

Speakers as Microphones

A loudspeaker, a transducer that turns an electrical signal into sound waves, is the functional opposite of a microphone. Since a conventional speaker is constructed much like a dynamic microphone (with a diaphragm, coil and magnet), speakers can actually work "in reverse" as microphones. The resulting signal typically offers reduced quality including limited high-end frequency response and poor sensitivity. In practical use, speakers are sometimes used as microphones in applications where high quality and sensitivity are not needed such as intercoms, walkie-talkies or video game voice chat peripherals, or when conventional microphones are in short supply.

However, there is at least one practical application that exploits those weaknesses: the use of a medium-size woofer placed closely in front of a "kick drum" (bass drum) in a drum set to act as a microphone. A commercial product example is the Yamaha Subkick, a 6.5-inch (170 mm) woofer shock-mounted into a 10" drum shell used in front of kick drums. Since a relatively massive membrane is unable to transduce high frequencies while being capable of tolerating strong low-frequency transients, the speaker is often ideal for picking up the kick drum while reducing bleed from the nearby cymbals and snare drums. Less commonly, microphones themselves can be used as speakers, but due to their low power handling and small transducer sizes, a tweeter is the most practical application. One instance of such an application was the STC microphone-derived 4001 super-tweeter, which was successfully used in a number of high quality loudspeaker systems from the late 1960s to the mid-70s.

Capsule Design and Directivity

The inner elements of a microphone are the primary source of differences in directivity. A pressure microphone uses a diaphragm between a fixed internal volume of air and the environment, and responds uniformly to pressure from all directions, so it is said to be omnidirectional. A pressure-gradient microphone uses a diaphragm that is at least partially open on both sides. The pressure difference between the two sides produces its directional characteristics. Other elements such as the external shape of the microphone and external devices such as interference tubes can also alter a microphone's directional response. A pure pressure-gradient microphone is equally sensitive to sounds arriving from front or back, but insensitive to sounds arriving from the side because sound arriving at the front and back at the same time creates no gradient between the two. The characteristic directional pattern of a pure pressure-gradient microphone is like a figure-8. Other polar patterns are derived by creating a capsule that combines these two effects in different ways. The cardioid, for instance, features a partially closed backside, so its response is a combination of pressure and pressure-gradient characteristics.

Microphone Polar Patterns

A microphone's directionality or polar pattern indicates how sensitive it is to sounds arriving at different angles about its central axis. The polar patterns illustrated above represent the locus of points that produce the same signal level output in the microphone if a given sound pressure level (SPL) is generated from that point. How the physical body of the microphone is oriented relative to the diagrams depends on the microphone design. For large-membrane microphones such as in the Oktava (pictured above), the upward direction in the polar diagram is usually perpendicular to the microphone body, commonly known as "side fire" or "side address". For small diaphragm microphones such as the Shure (also pictured above), it usually extends from the axis of the microphone commonly known as "end fire" or "top/end address".

Some microphone designs combine several principles in creating the desired polar pattern. This ranges from shielding (meaning diffraction/dissipation/absorption) by the housing itself to electronically combining dual membranes.

Omnidirectional

An omnidirectional (or nondirectional) microphone's response is generally considered to be a perfect sphere in three dimensions. In the real world, this is not the case. As with directional microphones, the polar pattern for an "omnidirectional" microphone is a function of frequency. The body of the microphone is not infinitely small and, as a consequence, it tends to get in its own way with respect to sounds arriving from the rear, causing a slight flattening of the polar response. This flattening increases as the diameter of the microphone (assuming it's cylindrical) reaches the wavelength of the frequency in question. Therefore, the smallest diameter microphone gives the best omnidirectional characteristics at high frequencies.

The wavelength of sound at 10 kHz is 1.4" (3.5 cm). The smallest measuring microphones are often 1/4" (6 mm) in diameter, which practically eliminates directionality even up to the highest frequencies. Omnidirectional microphones, unlike cardioids, do not employ resonant cavities as delays, and so can be considered the "purest" microphones in terms of low coloration; they add very little to the original sound. Being pressure-sensitive they can also have a very flat low-frequency response down to 20 Hz or below. Pressure-sensitive microphones also respond much less to wind noise and plosives than directional (velocity sensitive) microphones.

An example of a nondirectional microphone is the round black *eight ball*.

Unidirectional

A unidirectional microphone is primarily sensitive to sounds from only one direction. The diagram above illustrates a number of these patterns. The microphone faces upwards in each diagram. The sound intensity for a particular frequency is plotted for angles radially from 0 to 360°. (Professional diagrams show these scales and include multiple plots at different frequencies. The diagrams given here provide only an overview of typical pattern shapes, and their names.)

Cardioid, Hypercardioid, Supercardioid

The most common unidirectional microphone is a cardioid microphone, so named because the sensitivity pattern is "heart-shaped", i.e. a cardioid. The cardioid family of microphones are commonly used as vocal or speech microphones, since they are good at rejecting sounds from other directions. In three dimensions, the cardioid is shaped like an apple centred around the microphone which is the "stem" of the apple. The cardioid response reduces pickup from the side and rear, helping to avoid feedback from the monitors. Since these directional transducer microphones achieve their patterns by sensing pressure gradient, putting them very close to the sound source (at distances of a few centimeters) results in a bass boost due to the increased gradient. This is known as the proximity effect. The SM58 has been the most commonly used microphone for live vocals for more than 50 years demonstrating the importance and popularity of cardioid mics.

University Sound US664A dynamic supercardioid microphone

A cardioid microphone is effectively a superposition of an omnidirectional and a figure-8 microphone; for sound waves coming from the back, the negative signal from the figure-8 cancels the positive signal from the omnidirectional element, whereas for sound waves coming from the front, the two add to each other. A hyper-cardioid microphone is similar, but with a slightly larger figure-8 contribution leading to a tighter area of front sensitivity and a smaller lobe of rear sensitivity. A super-cardioid microphone is similar to a hyper-cardioid, except there is more front pickup and less rear pickup. While any pattern between omni and figure 8 is possible by adjusting their mix, common definitions state that a hypercardioid is produced by combining them at a 3:1 ratio, producing nulls at 109.5°, while supercardioid is produced with a 5:3 ratio, with nulls at 126.9°.

Bi-directional

"Figure 8" or bi-directional microphones receive sound equally from both the front and back of the element. Most ribbon microphones are of this pattern. In principle they do not respond to sound pressure at all, only to the *change* in pressure between front and back; since sound arriving from the side reaches front and back equally there is no difference in pressure and therefore no sensitivity to sound from that direction. In more mathematical terms, while omnidirectional microphones are scalar transducers responding to pressure from any direction, bi-directional microphones are vector transducers responding to the gradient along an axis normal to the plane of the diaphragm. This also has the effect of inverting the output polarity for sounds arriving from the back side.

Shotgun and Parabolic Microphones

An Audio-Technica shotgun microphone

Shotgun microphones are the most highly directional of simple first-order unidirectional types. At low frequencies they have the classic polar response of a hypercardioid but at medium and higher frequencies an interference tube gives them an increased forward response. This is achieved by a process of cancellation of off-axis waves entering the longitudinal array of slots. A consequence of this technique is the presence of some rear lobes that vary in level and angle with frequency, and can cause some coloration effects. Due to the narrowness of their forward sensitivity, shotgun microphones are commonly used on television and film sets, in stadiums, and for field recording of wildlife.

The interference tube of a shotgun microphone. The capsule is at the base of the tube.

A Sony parabolic reflector, without a microphone. The microphone would face the reflector surface and sound captured by the reflector would bounce towards the microphone.

Boundary or "PZM"

Several approaches have been developed for effectively using a microphone in less-than-ideal acoustic spaces, which often suffer from excessive reflections from one or more of the surfaces (boundaries) that make up the space. If the microphone is placed in, or very close to, one of these boundaries, the reflections from that surface have the same timing as the direct sound, thus giving the microphone a hemispherical polar pattern and improved intelligibility. Initially this was done by placing an ordinary microphone adjacent to the surface, sometimes in a block of acoustically transparent foam. Sound engineers Ed Long and Ron Wickersham developed the concept of placing the diaphragm parallel to and facing the boundary. While the patent has expired, "Pressure Zone Microphone" and "PZM" are still active trademarks of Crown International, and the generic term "boundary microphone" is preferred. While a boundary microphone was initially implemented using an omnidirectional element, it is also possible to mount a directional microphone close enough to the surface to gain some of the benefits of this technique while retaining the directional properties of the element. Crown's trademark on this approach is "Phase Coherent Cardioid" or "PCC," but there are other makers who employ this technique as well.

Application-specific Designs

A lavalier microphone is made for hands-free operation. These small microphones are worn on the body. Originally, they were held in place with a lanyard worn around the neck, but more often they are fastened to clothing with a clip, pin, tape or magnet. The lavalier cord may be hidden by clothes and either run to an RF transmitter in a pocket or clipped to a belt (for mobile use), or run directly to the mixer (for stationary applications).

A wireless microphone transmits the audio as a radio or optical signal rather than via a cable. It usually sends its signal using a small FM radio transmitter to a nearby receiver connected to the sound system, but it can also use infrared waves if the transmitter and receiver are within sight of each other.

A contact microphone picks up vibrations directly from a solid surface or object, as opposed to sound vibrations carried through air. One use for this is to detect sounds of a very low level, such as those from small objects or insects. The microphone commonly consists of a magnetic (moving coil) transducer, contact plate and contact pin. The contact plate is placed directly on the vibrating part of a musical instrument or other surface, and the contact pin transfers vibrations to the coil. Contact microphones have been used to pick up the sound of a snail's heartbeat and the footsteps of ants. A portable version of this microphone has recently been developed. A throat microphone is a variant of the contact microphone that picks up speech directly from a person's throat, which it is strapped to. This lets the device be used in areas with ambient sounds that would otherwise make the speaker inaudible.

A parabolic microphone uses a parabolic reflector to collect and focus sound waves onto a microphone receiver, in much the same way that a parabolic antenna (e.g. satellite dish) does with radio waves. Typical uses of this microphone, which has unusually focused front sensitivity and can pick up sounds from many meters away, include nature recording, outdoor sporting events, eavesdropping, law enforcement, and even espionage. Parabolic microphones are not typically used for standard recording applications, because they tend to have poor low-frequency response as a side effect of their design.

A stereo microphone integrates two microphones in one unit to produce a stereophonic signal. A stereo microphone is often used for broadcast applications or field recording where it would be impractical to configure two separate condenser microphones in a classic X-Y configuration for stereophonic recording. Some such microphones have an adjustable angle of coverage between the two channels.

A noise-canceling microphone is a highly directional design intended for noisy environments. One such use is in aircraft cockpits where they are normally installed as boom microphones on head-sets. Another use is in live event support on loud concert stages for vocalists involved with live performances. Many noise-canceling microphones combine signals received from two diaphragms that are in opposite electrical polarity or are processed electronically. In dual diaphragm designs, the main diaphragm is mounted closest to the intended source and the second is positioned farther away from the source so that it can pick up environmental sounds to be subtracted from the main diaphragm's signal. After the two signals have been combined, sounds other than the intended source are greatly reduced, substantially increasing intelligibility. Other noise-canceling designs use one diaphragm that is affected by ports open to the sides and rear of the microphone, with the sum being a 16 dB rejection of sounds that are farther away. One noise-canceling headset design using a single diaphragm has been used prominently by vocal artists such as Garth Brooks and Janet Jackson. A few noise-canceling microphones are throat microphones.

Powering

Microphones containing active circuitry, such as most condenser microphones, require power to operate the active components. The first of these used vacuum-tube circuits with a separate power supply unit, using a multi-pin cable and connector. With the advent of solid-state amplification, the power requirements were greatly reduced and it became practical to use the same cable conductors and connector for audio and power. During the 1960s several powering methods were developed, mainly in Europe. The two dominant methods were initially defined in German DIN 45595 as de:Tonaderspeisung or T-power and DIN 45596 for phantom power. Since the 1980s, phantom power has become much more common, because the same input may be used for both powered and unpowered microphones. In consumer electronics such as DSLRs and camcorders, "plug-in power" is more common, for microphones using a 3.5 mm phone plug connector. Phantom, T-power and plug-in power are described in international standard IEC 61938.

Connectors

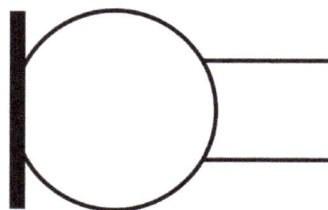

Electronic symbol for a microphone

The most common connectors used by microphones are:

- Male XLR connector on professional microphones

- ¼ inch (sometimes referred to as 6.3 mm) phone connector on less expensive musician's microphones, using an unbalanced 1/4 inch (6.3 mm) TS phone connector. Harmonica microphones commonly use a high impedance 1/4 inch (6.3 mm) TS connection to be run through guitar amplifiers.

- 3.5 mm (sometimes referred to as 1/8 inch mini) stereo (sometimes wired as mono) mini phone plug on prosumer camera, recorder and computer microphones.

A microphone with a USB connector, made by Blue Microphones

Some microphones use other connectors, such as a 5-pin XLR, or mini XLR for connection to portable equipment. Some lavalier (or "lapel", from the days of attaching the microphone to the news reporters suit lapel) microphones use a proprietary connector for connection to a wireless transmitter, such as a radio pack. Since 2005, professional-quality microphones with USB connections have begun to appear, designed for direct recording into computer-based software.

Impedance-matching

Microphones have an electrical characteristic called impedance, measured in ohms (Ω), that depends on the design. In passive microphones, this value describes the electrical resistance of the magnet coil (or similar mechanism). In active microphones, this value describes the output resistance of the amplifier circuitry. Typically, the *rated impedance* is stated. Low impedance is considered under 600 Ω. Medium impedance is considered between 600 Ω and 10 kΩ. High impedance is above 10 kΩ. Owing to their built-in amplifier, condenser microphones typically have an output impedance between 50 and 200 Ω.

The output of a given microphone delivers the same power whether it is low or high impedance. If a microphone is made in high and low impedance versions, the high impedance version has a higher output voltage for a given sound pressure input, and is suitable for use with vacuum-tube guitar amplifiers, for instance, which have a high input impedance and require a relatively high signal input voltage to overcome the tubes' inherent noise. Most professional microphones are low impedance, about 200 Ω or lower. Professional vacuum-tube sound equipment incorporates a transformer that steps up the impedance of the microphone circuit to the high impedance and voltage needed to drive the input tube. External matching transformers are also available that can be used in-line between a low impedance microphone and a high impedance input.

Low-impedance microphones are preferred over high impedance for two reasons: one is that using a high-impedance microphone with a long cable results in high frequency signal loss due to cable capacitance, which forms a low-pass filter with the microphone output impedance. The other is that long high-impedance cables tend to pick up more hum (and possibly radio-frequency interference (RFI) as well). Nothing is damaged if the impedance between microphone and other equipment is mismatched; the worst that happens is a reduction in signal or change in frequency response.

Some microphones are designed *not* to have their impedance matched by the load they are connected to. Doing so can alter their frequency response and cause distortion, especially at high sound pressure levels. Certain ribbon and dynamic microphones are exceptions, due to the designers' assumption of a certain load impedance being part of the internal electro-acoustical damping circuit of the microphone.

Digital Microphone Interface

The AES42 standard, published by the Audio Engineering Society, defines a digital interface for microphones. Microphones conforming to this standard directly output a digital audio stream through an XLR or XLD male connector, rather than producing an analog output. Digital microphones may be used either with new equipment with appropriate input connections that conform to the AES42 standard, or else via a suitable interface box. Studio-quality microphones that operate in accordance with the AES42 standard are now available from a number of microphone manufacturers.

Neumann D-01 digital microphone and Neumann DMI-8 8-channel USB Digital Microphone Interface

Measurements and Specifications

Because of differences in their construction, microphones have their own characteristic responses to sound. This difference in response produces non-uniform phase and frequency responses. In addition, microphones are not uniformly sensitive to sound pressure, and can accept differing levels without distorting. Although for scientific applications microphones with a more uniform response are desirable, this is often not the case for music recording, as the non-uniform response of a microphone can produce a desirable coloration of the sound. There is an international standard for microphone specifications, but few manufacturers adhere to it. As a result, comparison of published data from different manufacturers is difficult because different measurement techniques are used. The Microphone Data Website has collated the technical specifications complete with pictures, response curves and technical data from the microphone manufacturers for every currently listed microphone, and even a few obsolete models, and shows the data for them all in one common format for ease of comparison.. Caution should be used in drawing any solid conclusions from this or any other published data, however, unless it is known that the manufacturer has supplied specifications in accordance with IEC 60268-4.

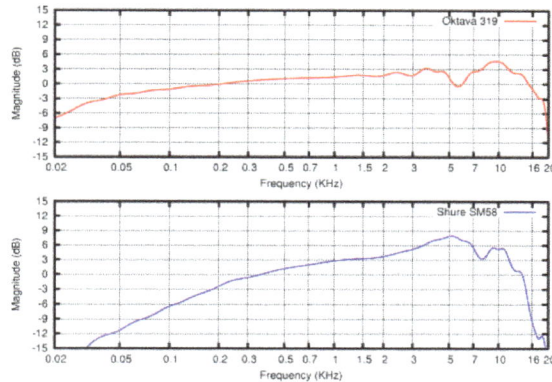

A comparison of the far field on-axis frequency response of the Oktava 319 and the Shure SM58

A frequency response diagram plots the microphone sensitivity in decibels over a range of frequencies (typically 20 Hz to 20 kHz), generally for perfectly on-axis sound (sound arriving at 0° to the capsule). Frequency response may be less informatively stated textually like so: "30 Hz–16 kHz ±3 dB". This is interpreted as meaning a nearly flat, linear, plot between the stated frequencies, with variations in amplitude of no more than plus or minus 3 dB. However, one cannot determine from this information how *smooth* the variations are, nor in what parts of the spectrum they occur. Note that commonly made statements such as "20 Hz–20 kHz" are meaningless without a decibel measure of tolerance. Directional microphones' frequency response varies greatly with distance from the sound source, and with the geometry of the sound source. IEC 60268-4 specifies that frequency response should be measured in *plane progressive wave* conditions (very far away from the source) but this is seldom practical. *Close talking* microphones may be measured with different sound sources and distances, but there is no standard and therefore no way to compare data from different models unless the measurement technique is described.

The self-noise or equivalent input noise level is the sound level that creates the same output voltage as the microphone does in the absence of sound. This represents the lowest point of the microphone's dynamic range, and is particularly important should you wish to record sounds that are quiet. The measure is often stated in dB(A), which is the equivalent loudness of the noise on a decibel scale frequency-weighted for how the ear hears, for example: "15 dBA SPL" (SPL means sound pressure level relative to 20 micropascals). The lower the number the better. Some microphone manufacturers state the noise level using ITU-R 468 noise weighting, which more accurately represents the way we hear noise, but gives a figure some 11–14 dB higher. A quiet microphone typically measures 20 dBA SPL or 32 dB SPL 468-weighted. Very quiet microphones have existed for years for special applications, such the Brüel & Kjaer 4179, with a noise level around 0 dB SPL. Recently some microphones with low noise specifications have been introduced in the studio/entertainment market, such as models from Neumann and Røde that advertise noise levels between 5–7 dBA. Typically this is achieved by altering the frequency response of the capsule and electronics to result in lower noise within the A-weighting curve while broadband noise may be increased.

The maximum SPL the microphone can accept is measured for particular values of total harmonic distortion (THD), typically 0.5%. This amount of distortion is generally inaudible, so one can safely use the microphone at this SPL without harming the recording. Example: "142 dB SPL peak (at 0.5% THD)". The higher the value, the better, although microphones with a very high maximum SPL also have a higher self-noise.

The clipping level is an important indicator of maximum usable level, as the 1% THD figure usually quoted under max SPL is really a very mild level of distortion, quite inaudible especially on brief high peaks. Clipping is much more audible. For some microphones the clipping level may be much higher than the max SPL.

The dynamic range of a microphone is the difference in SPL between the noise floor and the maximum SPL. If stated on its own, for example "120 dB", it conveys significantly less information than having the self-noise and maximum SPL figures individually.

Sensitivity indicates how well the microphone converts acoustic pressure to output voltage. A high sensitivity microphone creates more voltage and so needs less amplification at the mixer or recording device. This is a practical concern but is not directly an indication of the microphone's quality, and in fact the term sensitivity is something of a misnomer, "transduction gain" being perhaps more meaningful, (or just "output level") because true sensitivity is generally set by the noise floor, and too much "sensitivity" in terms of output level compromises the clipping level. There are two common measures. The (preferred) international standard is made in millivolts per pascal at 1 kHz. A higher value indicates greater sensitivity. The older American method is referred to a 1 V/ Pa standard and measured in plain decibels, resulting in a negative value. Again, a higher value indicates greater sensitivity, so –60 dB is more sensitive than –70 dB.

Measurement Microphones

Some microphones are intended for testing speakers, measuring noise levels and otherwise quantifying an acoustic experience. These are calibrated transducers and are usually supplied with a calibration certificate that states absolute sensitivity against frequency. The quality of measurement microphones is often referred to using the designations "Class 1," "Type 2" etc., which are references not to microphone specifications but to sound level meters. A more comprehensive standard for the description of measurement microphone performance was recently adopted.

Measurement microphones are generally scalar sensors of pressure; they exhibit an omnidirectional response, limited only by the scattering profile of their physical dimensions. Sound intensity or sound power measurements require pressure-gradient measurements, which are typically made using arrays of at least two microphones, or with hot-wire anemometers.

Microphone Calibration

To take a scientific measurement with a microphone, its precise sensitivity must be known (in volts per pascal). Since this may change over the lifetime of the device, it is necessary to regularly calibrate measurement microphones. This service is offered by some microphone manufacturers and by independent certified testing labs. All microphone calibration is ultimately traceable to primary standards at a national measurement institute such as NPL in the UK, PTB in Germany and NIST in the United States, which most commonly calibrate using the reciprocity primary standard. Measurement microphones calibrated using this method can then be used to calibrate other microphones using comparison calibration techniques.

Depending on the application, measurement microphones must be tested periodically (every year or several months, typically) and after any potentially damaging event, such as being dropped (most such microphones come in foam-padded cases to reduce this risk) or exposed to sounds beyond the acceptable level.

Microphone Array and Array Microphones

A microphone array is any number of microphones operating in tandem. There are many applications:

- Systems for extracting voice input from ambient noise (notably telephones, speech recognition systems, hearing aids)

- Surround sound and related technologies

- Locating objects by sound: acoustic source localization, *e.g.*, military use to locate the source(s) of artillery fire. Aircraft location and tracking.

- High fidelity original recordings

- 3D spatial beamforming for localized acoustic detection of subcutaneous sounds

Typically, an array is made up of omnidirectional microphones distributed about the perimeter of a space, linked to a computer that records and interprets the results into a coherent form.

Microphone Windscreens

Windscreens (or windshields – the terms are interchangeable) provide a method of reducing the effect of wind on microphones. While pop-screens give protection from unidirectional blasts, foam "hats" shield wind into the grille from all directions, and blimps / zeppelins / baskets entirely enclose the microphone and protect its body as well. This last point is important because, given the extreme low frequency content of wind noise, vibration induced in the housing of the microphone can contribute substantially to the noise output.

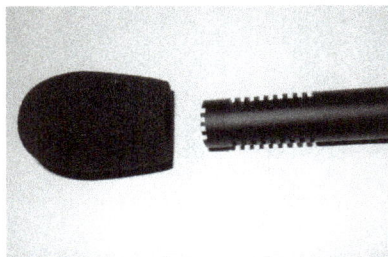

Microphone with its windscreen removed.

The shielding material used – wire gauze, fabric or foam – is designed to have a significant acoustic impedance. The relatively low particle-velocity air pressure changes that constitute sound waves can pass through with minimal attenuation, but higher particle-velocity wind is impeded to a far greater extent. Increasing the thickness of the material improves wind attenuation but also begins to compromise high frequency audio content. This limits the practical size of simple foam screens. While foams and wire meshes can be partly or wholly self-supporting, soft fabrics and gauzes require stretching on frames, or laminating with coarser structural elements.

Since all wind noise is generated at the first surface the air hits, the greater the spacing between shield periphery and microphone capsule, the greater the noise attenuation. For an approximately spherical shield, attenuation increases by (approximately) the cube of that distance. Thus larger shields are always much more efficient than smaller ones. With full basket windshields there is an additional pressure chamber effect, first explained by Joerg Wuttke, which, for two-port (pressure gradient) microphones, allows the shield/microphone combination to act as a high-pass acoustic filter.

Since turbulence at a surface is the source of wind noise, reducing gross turbulence can add to noise reduction. Both aerodynamically smooth surfaces, and ones that prevent powerful vortices being generated, have been used successfully. Historically, artificial fur has proved very useful for this purpose since the fibres produce micro-turbulence and absorb energy silently. If not matted by wind and rain, the fur fibres are very transparent acoustically, but the woven or knitted backing can give significant attenuation. As a material it suffers from being difficult to manufacture with consistency, and to keep in pristine condition on location. Thus there is an interest (DPA 5100, Rycote Cyclone) to move away from its use.

In the studio and on stage, pop-screens and foam shields can be useful for reasons of hygiene, and protecting microphones from spittle and sweat. They can also be useful coloured idents. On location the basket shield can contain a suspension system to isolate the microphone from shock and handling noise.

Various microphone covers

Stating the efficiency of wind noise reduction is an inexact science, since the effect varies enormously with frequency, and hence with the bandwidth of the microphone and audio channel. At very low frequencies (10–100 Hz) where massive wind energy exists, reductions are important to avoid overloading of the audio chain – particularly the early stages. This can produce the typical "wumping" sound associated with wind, which is often syllabic muting of the audio due to LF peak limiting. At higher frequencies – 200 Hz to ~3 kHz – the aural sensitivity curve allows us to hear the effect of wind as an addition to the normal noise floor, even though it has a far lower energy content. Simple shields may allow the wind noise to be 10 dB less apparent; better ones can achieve nearer to a 50 dB reduction. However the acoustic transparency, particularly at HF, should also be indicated, since a very high level of wind attenuation could be associated with very muffled audio.

Two recordings being made—a *blimp* is being used on the left. An open-cell foam windscreen is being used on the right.

Hydrophone

A hydrophone is a microphone designed to be used underwater for recording or listening to underwater sound. Most hydrophones are based on a piezoelectric transducer that generates electricity when subjected to a pressure change. Such piezoelectric materials, or transducers, can convert a sound signal into an electrical signal since sound is a pressure wave. Some transducers can also serve as a projector, but not all have this capability, and some may be destroyed if used in such a manner.

A hydrophone can "listen" to sound in air but will be less sensitive due to its design as having a good acoustic impedance match to water, which is a denser fluid than air. Likewise, a microphone can be buried in the ground, or immersed in water if it is put in a waterproof container, but will give similarly poor performance due to the similarly bad acoustic impedance match.

A hydrophone

History

The earliest widely used design was the Fessenden oscillator, an electrodynamically driven clamped-edge circular plate transducer (not actually an oscillator) operating at 500, 1000, and later 3000 Hz. It was originally marketed as an underwater telegraph, rather than as sonar, but was later very successful, its Canadian inventor, Reginald Fessenden, was awarded the "Scientific American Magazine Gold Medal of Safety" in 1929 from the American Museum of Safety, an organization for ship captains; some were still in use during World War II.

A hydrophone being lowered into the North Atlantic

Ernest Rutherford, in England, led pioneer research in hydrophones using piezoelectric devices, and his only patent was for a hydrophone device. The acoustic impedance of piezoelectric materials facilitated their use as underwater transducers. The piezoelectric hydrophone was used late in World War I, by convoy escorts detecting U-boats, greatly impacting the effectiveness of submarines.

From late in World War I until the introduction of active sonar, hydrophones were the sole method for submarines to detect targets while submerged, and remain useful today.

Directional Hydrophones

A small single cylindrical ceramic transducer can achieve near perfect omnidirectional reception. Directional hydrophones increase sensitivity from one direction using two basic techniques:

Focused Transducers

This device uses a single transducer element with a dish or conical-shaped sound reflector to focus the signals, in a similar manner to a reflecting telescope. This type of hydrophone can be produced from a low-cost omnidirectional type, but must be used while stationary, as the reflector impedes its movement through water. A new way to direct is to use a spherical body around the hydrophone. The advantage of directivity spheres is that the hydrophone can be moved within the water, ridding it of the interferences produced by a conical-shaped element

Arrays

Multiple hydrophones can be arranged in an array so that it will add the signals from the desired direction while subtracting signals from other directions. The array may be steered using a beamformer. Most commonly, hydrophones are arranged in a "line array" but may be in two- or three-dimensional arrangements.

SOSUS hydrophones, laid on the seabed and connected by underwater cables, were used, beginning in the 1950s, by the U.S. Navy to track movement of Soviet submarines during the Cold War along a line from Greenland, Iceland and the United Kingdom known as the GIUK gap. These are capable of clearly recording extremely low frequency infrasound, including many unexplained ocean sounds.

References

- To cite a book with a credited author Winer, Ethan (2013). "Part 3". The Audio Expert. New York and London: Focal Press. ISBN 978-0-240-82100-9.

- Ballou, Glen (2013). Handbook for Sound Engineers, 4th Ed. Taylor and Francis. p. 597. ISBN 1136122532.

- Talbot-Smith, Michael (2013). Audio Engineer's Reference Book. CRC Press. p. 2.52. ISBN 1136119744.

- Spanias, Andreas; Ted Painter; Venkatraman Atti (2007). Audio Signal Processing and Coding. Wiley-Interscience. ISBN 0-470-04196-X.

- Rumsey, Francis; McCormick, Tim (2009). Sound and recording (6th ed.). Oxford, UK: Focal Press. p. 81. ISBN 978-0-240-52163-3.

- Davis, Don; Carolyn Davis (1997). "Loudspeakers and Loudspeaker Arrays". Sound System Engineering (2 ed.). Focal Press. p. 350. ISBN 0-240-80305-1. Retrieved March 30, 2010. We often give lip service to the fact that audio allows its practitioners to engage in both art and science.

- Eargle, John; Chris Foreman (2002). Audio Engineering for Sound Reinforcement. Milwaukee: Hal Leonard Corporation. p. 66. ISBN 0-634-04355-2.

- "Glossary of Terms". Home Theater Design. ETS-eTech. p. 1. Archived from the original on July 23, 2012. Retrieved March 3, 2010.

- "Lee De Forest – (1873–1961)". Television International Magazine. 2011-01-17. Archived from the original on 2011-01-17. Retrieved Dec 4, 2013.

- "Kyocera piezoelectric film speaker delivers 180-degree sound to thin TVs and tablets (update: live photos)". August 29, 2013.

6

Principles and Applications of Transduction

The force that exists between electrically charged particles is termed as electromagnetic force. Electromagnetism studies the phenomena of electromagnetic force. Transduction involves the application of certain principles like electromagnetism, electrostatics, piezoelectricity and sound reinforcement system. This section seeks to explain the fundamentals of transduction.

Electromagnetism

Electromagnetism is a branch of physics which involves the study of the electromagnetic force, a type of physical interaction that occurs between electrically charged particles. The electromagnetic force usually exhibits electromagnetic fields, such as electric fields, magnetic fields, and light. The electromagnetic force is one of the four fundamental interactions (commonly called forces) in nature. The other three fundamental interactions are the strong interaction, the weak interaction, and gravitation.

Lightning is an electrostatic discharge that travels between two charged regions.

The word *electromagnetism* is a type of iron ore. Electromagnetic phenomena is defined in terms of the electromagnetic force, sometimes called the Lorentz force, which includes both electricity and magnetism as different manifestations of the same phenomenon.

The electromagnetic force plays a major role in determining the internal properties of most objects encountered in daily life. Ordinary matter takes its form as a result of intermolecular forces between individual atoms and molecules in matter, and are a manifestation of the electromagnetic force. Electrons are bound by the electromagnetic force to atomic nuclei, and their orbital shapes and their influence on nearby atoms with their electrons is described by quantum mechanics. The electromagnetic force governs the processes involved in chemistry, which arise from interactions between the electrons of neighboring atoms.

There are numerous mathematical descriptions of the electromagnetic field. In classical electro-dynamics, electric fields are described as electric potential and electric current. In Faraday's law, magnetic fields are associated with electromagnetic induction and magnetism, and Maxwell's equations describe how electric and magnetic fields are generated and altered by each other and by charges and currents.

The theoretical implications of electromagnetism, in particular the establishment of the speed of light based on properties of the "medium" of propagation (permeability and permittivity), led to the development of special relativity by Albert Einstein in 1905.

Although electromagnetism is considered one of the four fundamental forces, at high energy the weak force and electromagnetic force are unified as a single electroweak force. In the history of the universe, during the quark epoch the unified force broke into the two separate forces as the universe cooled.

History of The Theory

Originally, electricity and magnetism were thought of as two separate forces. This view changed, however, with the publication of James Clerk Maxwell's 1873 *A Treatise on Electricity and Magnetism* in which the interactions of positive and negative charges were shown to be mediated by one force. There are four main effects resulting from these interactions, all of which have been clearly demonstrated by experiments:

Hans Christian Ørsted.

1. Electric charges attract or repel one another with a force inversely proportional to the square of the distance between them: unlike charges attract, like ones repel.

2. Magnetic poles (or states of polarization at individual points) attract or repel one another in a manner similar to positive and negative charges and always exist as pairs: every north pole is yoked to a south pole.

3. An electric current inside a wire creates a corresponding circumferential magnetic field outside the wire. Its direction (clockwise or counter-clockwise) depends on the direction of the current in the wire.

4. A current is induced in a loop of wire when it is moved toward or away from a magnetic field, or a magnet is moved towards or away from it; the direction of current depends on that of the movement.

André-Marie Ampère

While preparing for an evening lecture on 21 April 1820, Hans Christian Ørsted made a surprising observation. As he was setting up his materials, he noticed a compass needle deflected away from magnetic north when the electric current from the battery he was using was switched on and off. This deflection convinced him that magnetic fields radiate from all sides of a wire carrying an electric current, just as light and heat do, and that it confirmed a direct relationship between electricity and magnetism.

Michael Faraday

At the time of discovery, Ørsted did not suggest any satisfactory explanation of the phenomenon, nor did he try to represent the phenomenon in a mathematical framework. However, three months later he began more intensive investigations. Soon thereafter he published his findings, proving that an electric current produces a magnetic field as it flows through a wire. The CGS unit of magnetic induction (oersted) is named in honor of his contributions to the field of electromagnetism.

James Clerk Maxwell

His findings resulted in intensive research throughout the scientific community in electrodynamics. They influenced French physicist André-Marie Ampère's developments of a single mathematical form to represent the magnetic forces between current-carrying conductors. Ørsted's discovery also represented a major step toward a unified concept of energy.

This unification, which was observed by Michael Faraday, extended by James Clerk Maxwell, and partially reformulated by Oliver Heaviside and Heinrich Hertz, is one of the key accomplishments of 19th century mathematical physics. It had far-reaching consequences, one of which was the understanding of the nature of light. Unlike what was proposed by electromagnetic theory of that time, light and other electromagnetic waves are at the present seen as taking the form of quantized, self-propagating oscillatory electromagnetic field disturbances called photons. Different frequencies of oscillation give rise to the different forms of electromagnetic radiation, from radio waves at the lowest frequencies, to visible light at intermediate frequencies, to gamma rays at the highest frequencies.

Ørsted was not the only person to examine the relationship between electricity and magnetism. In 1802, Gian Domenico Romagnosi, an Italian legal scholar, deflected a magnetic needle using electrostatic charges. Actually, no galvanic current existed in the setup and hence no electromagnetism was present. An account of the discovery was published in 1802 in an Italian newspaper, but it was largely overlooked by the contemporary scientific community.

Fundamental Forces

The electromagnetic force is one of the four known fundamental forces. The other fundamental forces are:

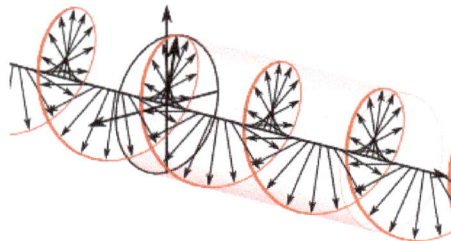

Representation of the electric field vector of a wave of circularly polarized electromagnetic radiation.

- the weak nuclear force, which binds to all known particles in the Standard Model, and causes certain forms of radioactive decay. (In particle physics though, the electroweak interaction is the unified description of two of the four known fundamental interactions of nature: electromagnetism and the weak interaction);

- the strong nuclear force, which binds quarks to form nucleons, and binds nucleons to form nuclei

- the gravitational force.

All other forces (e.g., friction, contact forces) are derived from these four fundamental forces (including momentum which is carried by the movement of particles).

The electromagnetic force is the one responsible for practically all the phenomena one encounters in daily life above the nuclear scale, with the exception of gravity. Roughly speaking, all the forces involved in interactions between atoms can be explained by the electromagnetic force acting between the electrically charged atomic nuclei and electrons of the atoms. Electromagnetic forces also explain how these particles carry momentum by their movement. This includes the forces we experience in "pushing" or "pulling" ordinary material objects, which result from the intermolecular forces that act between the individual molecules in our bodies and those in the objects. The electromagnetic force is also involved in all forms of chemical phenomena.

A necessary part of understanding the intra-atomic and intermolecular forces is the effective force generated by the momentum of the electrons' movement, such that as electrons move between interacting atoms they carry momentum with them. As a collection of electrons becomes more confined, their minimum momentum necessarily increases due to the Pauli exclusion principle. The behaviour of matter at the molecular scale including its density is determined by the balance between the electromagnetic force and the force generated by the exchange of momentum carried by the electrons themselves.

Classical Electrodynamics

The scientist William Gilbert proposed, in his *De Magnete* (1600), that electricity and magnetism, while both capable of causing attraction and repulsion of objects, were distinct effects. Mariners had noticed that lightning strikes had the ability to disturb a compass needle, but the link between lightning and electricity was not confirmed until Benjamin Franklin's proposed experiments in 1752. One of the first to discover and publish a link between man-made electric current and magnetism was Romagnosi, who in 1802 noticed that connecting a wire across a voltaic pile deflected a nearby compass needle. However, the effect did not become widely known until 1820, when Ørsted performed a similar experiment. Ørsted's work influenced Ampère to produce a theory of electromagnetism that set the subject on a mathematical foundation.

A theory of electromagnetism, known as classical electromagnetism, was developed by various physicists over the course of the 19th century, culminating in the work of James Clerk Maxwell, who unified the preceding developments into a single theory and discovered the electromagnetic nature of light. In classical electromagnetism, the behavior of the electromagnetic field is described by a set of equations known as Maxwell's equations, and the electromagnetic force is given by the Lorentz force law.

One of the peculiarities of classical electromagnetism is that it is difficult to reconcile with classi-cal mechanics, but it is compatible with special relativity. According to Maxwell's equations, the speed of light in a vacuum is a universal constant that is dependent only on the electrical permit-tivity and magnetic permeability of free space. This violates Galilean invariance, a long-standing cornerstone of classical mechanics. One way to reconcile the two theories (electromagnetism and classical mechanics) is to assume the existence of a luminiferous aether through which the light propagates. However, subsequent experimental efforts failed to detect the presence of the aether. After important contributions of Hendrik Lorentz and Henri Poincaré, in 1905, Albert Einstein solved the problem with the introduction of special relativity, which replaced classical kinematics with a new theory of kinematics compatible with classical electromagnetism.

In addition, relativity theory implies that in moving frames of reference a magnetic field transforms to a field with a nonzero electric component and conversely, a moving electric field transforms to a nonzero magnetic component, thus firmly showing that the phenomena are two sides of the same coin. Hence the term "electromagnetism".

Quantum Mechanics

Photoelectric Effect

In a second paper published in 1905, Albert Einstein undermined the very foundations of classical electromagnetism. In his theory of the photoelectric effect, for which he won the Nobel prize in physics, he posited that light could exist in discrete particle-like quantities, which later came to be called photons. Einstein's theory of the photoelectric effect extended the insights that appeared in the solution of the ultraviolet catastrophe presented by Max Planck in 1900 and who coined the term "quanta" . In his work, Planck showed that hot objects emit electromagnetic radiation in discrete packets ("quanta"), which leads to a finite total energy emitted as black body radiation. Both of these results were in direct contradiction with the classical view of light as a continuous wave. Planck's and Einstein's theories were progenitors of quantum mechanics, which, when formulated in 1925, necessitated the invention of a quantum theory of electromagnetism. This theory, completed in the 1940s-1950s, is known as quantum electrodynamics (or "QED"), and, in situations where perturbation theory can be applied, is one of the most accurate theories known to physics.

Quantum Electrodynamics

All electromagnetic phenomena can be explained in terms of quantum mechanics, specifically by quantum electrodynamics (which includes classical electrodynamics as a limiting case) and this accounts for almost all physical phenomena observable to the unaided human senses, including electromagnetic radiation (light), all of chemistry, most of mechanics (excepting gravitation), and, of course, magnetism and electricity.

Electroweak Interaction

The electroweak interaction is the unified field theory description of two of the four known fundamental interactions of nature: electromagnetism and the weak interaction. Although these two forces appear very different at everyday low energies, the theory models them as two different aspects of the same force. At energies greater than 100 GeV, called the unification energy, the two forces merge into a single electroweak force. Thus when the universe was hot enough (approximately 10^{15} K, a temperature that was exceeded until shortly after the Big Bang) the electromagnetic force and weak force were merged into the electroweak force. With the cooling of the universe, during the electroweak epoch, the electroweak force separated from the strong force. Following that it was still too hot for quarks to combine into hadrons and they moved about freely.

Quantities and Units

Electromagnetic units are part of a system of electrical units based primarily upon the magnetic properties of electric currents, the fundamental SI unit being the ampere. The units are:

- ampere (electric current)
- coulomb (electric charge)
- farad (capacitance)
- henry (inductance)
- ohm (resistance)
- siemens (conductance)
- tesla (magnetic flux density)
- volt (electric potential)
- watt (power)
- weber (magnetic flux)

In the electromagnetic cgs system, electric current is a fundamental quantity defined via Ampère's law and takes the permeability as a dimensionless quantity (relative permeability) whose value in a vacuum is unity. As a consequence, the square of the speed of light appears explicitly in some of the equations interrelating quantities in this system.

SI electromagnetism units				
Symbol	Name of Quantity	Derived Units	Unit	Base Units
I	electric current	ampere (SI base unit)	A	A (= W/V = C/s)
Q	electric charge	coulomb	C	A.s
$U, \Delta V, \Delta\varphi; E$	potential difference; electromotive force	volt	V	$kg.m^2.s^{-3}.A^{-1}$ (= J/C)
$R; Z; X$	electric resistance; impedance; reactance	ohm	Ω	$kg.m^2.s^{-3}.A^{-2}$ (= V/A)
ρ	resistivity	ohm metre	Ω.m	$kg.m^3.s^{-3}.A^{-2}$
P	electric power	watt	W	$kg.m^2.s^{-3}$ (= V.A)
C	capacitance	farad	F	$kg^{-1}.m^{-2}.s^4.A^2$ (= C/V)
E	electric field strength	volt per metre	V/m	$kg.m.s^{-3}.A^{-1}$ (= N/C)

D	electric displacement field	coulomb per square metre	C/m²	A.s.m⁻²
ε	permittivity	farad per metre	F/m	kg⁻¹.m⁻³.s⁴.A²
χ_e	electric susceptibility	(dimensionless)	–	–
$G; Y; B$	conductance; admittance; susceptance	siemens	S	kg⁻¹.m⁻².s³.A² (= Ω⁻¹)
κ, γ, σ	conductivity	siemens per metre	S/m	kg⁻¹.m⁻³.s³.A²
B	magnetic flux density, magnetic induction	tesla	T	kg.s⁻².A⁻¹ (= Wb/m² = N.A⁻¹. m⁻¹)
	magnetic flux	weber	Wb	kg.m².s⁻².A⁻¹ (= V.s)
H	magnetic field strength	ampere per metre	A/m	A.m⁻¹
L, M	inductance	henry	H	kg.m².s⁻².A⁻² (= Wb/A = V.s/A)
μ	permeability	henry per metre	H/m	kg.m.s⁻².A⁻²
χ	magnetic susceptibility	(dimensionless)	–	–

Formulas for physical laws of electromagnetism (such as Maxwell's equations) need to be adjusted depending on what system of units one uses. This is because there is no one-to-one correspondence between electromagnetic units in SI and those in CGS, as is the case for mechanical units. Furthermore, within CGS, there are several plausible choices of electromagnetic units, leading to different unit "sub-systems", including Gaussian, "ESU", "EMU", and Heaviside–Lorentz. Among these choices, Gaussian units are the most common today, and in fact the phrase "CGS units" is often used to refer specifically to CGS-Gaussian units.

Electrostatics

Electrostatics is a branch of physics that deals with the phenomena and properties of stationary or slow-moving electric charges.

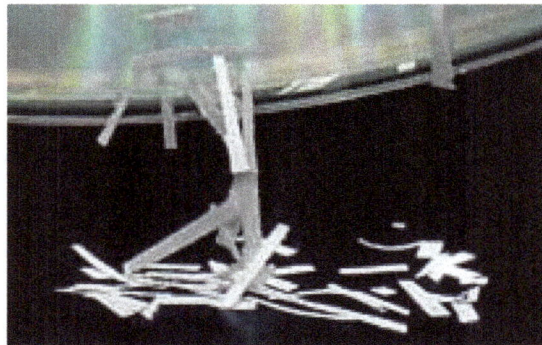

Paper strips attracted by a charged CD

Since classical physics, it has been known that some materials such as amber attract lightweight particles after rubbing. Electrostatic phenomena arise from the forces that electric charges exert on each other. Such forces are described by Coulomb's law. Even though electrostatically induced forces seem to be rather weak, the electrostatic force between e.g. an electron and a proton, that together make up a hydrogen atom, is about 36 orders of magnitude stronger than the gravitational force acting between them.

There are many examples of electrostatic phenomena, from those as simple as the attraction of the plastic wrap to your hand after you remove it from a package, and the attraction of paper to a charged scale, to the apparently spontaneous explosion of grain silos, the damage of electronic components during manufacturing, and photocopier & laser printer operation. Electrostatics involves the buildup of charge on the surface of objects due to contact with other surfaces. Although charge exchange happens whenever any two surfaces contact and separate, the effects of charge exchange are usually only noticed when at least one of the surfaces has a high resistance to electrical flow. This is because the charges that transfer are trapped there for a time long enough for their effects to be observed. These charges then remain on the object until they either bleed off to ground or are quickly neutralized by a discharge: e.g., the familiar phenomenon of a static 'shock' is caused by the neutralization of charge built up in the body from contact with insulated surfaces.

Coulomb's Law

We begin with the magnitude of the electrostatic force (in newtons) between two point charges q and Q (in coulombs). It is convenient to label one of these charges, q, as a test charge, and call Q a source charge. As we develop the theory, more source charges will be added. If r is the distance (in meters) between two charges, then the force is:

$$F = \frac{1}{4\pi\varepsilon_0}\frac{qQ}{r^2} = k_e\frac{qQ}{r^2},$$

where ε_0 is the vacuum permittivity, or permittivity of free space:

$$\varepsilon_0 = \frac{10^{-9}}{36\pi}\ \text{C}^2\ \text{N}^{-1}\ \text{m}^{-2} \approx 8.854\,187\,817 \times 10^{-12}\ \text{C}^2\ \text{N}^{-1}\ \text{m}^{-2}.$$

The SI units of ε_0 are equivalently $\text{A}^2\text{s}^4\ \text{kg}^{-1}\text{m}^{-3}$ or $\text{C}^2\text{N}^{-1}\text{m}^{-2}$ or F m^{-1}. Coulomb's constant is:

$$k_e \approx \frac{1}{4\pi\varepsilon_0} \approx 8.987\,551\,787 \times 10^9\ \text{N m}^2\ \text{C}^{-2}.$$

The use of ε_0 instead of k_0 in expressing Coulomb's Law is related to the fact that the force is inversely proportional to the surface area of a sphere with radius equal to the separation between the two charges.

A single proton has a charge of e, and the electron has a charge of $-e$, where,

$$e \approx 1.602\,176\,565 \times 10^{-19}\ \text{C}.$$

These physical constants (ε_0, k_0, e) are currently defined so that ε_0 and k_0 are exactly defined, and e is a measured quantity.

Electric Field

Electric field lines are useful for visualizing the electric field. Field lines begin on positive charge and terminate on negative charge. Electric field lines are parallel to the direction of the electric field, and the density of these field lines is a measure of the magnitude of the electric field at any given point. The electric field, $\vec{E}$, (in units of volts per meter) is a vector field that can be defined everywhere, except at the location of point charges (where it diverges to infinity). It is convenient to place a hypothetical test charge at a point (where no charges are present). By Coulomb's Law, this test charge will experience a force that can be used to define the electric field as follow

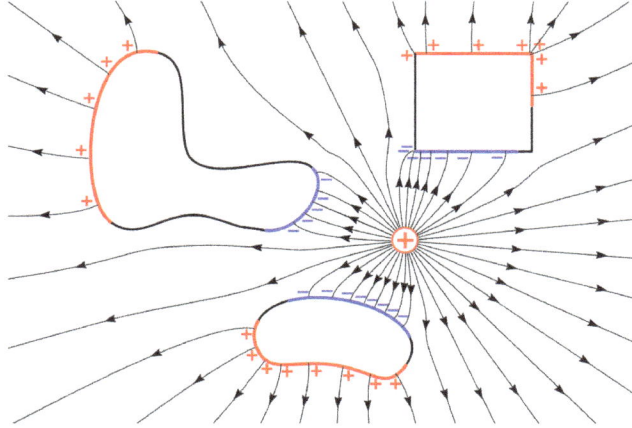

The electrostatic field *(lines with arrows)* of a nearby positive charge *(+)* causes the mobile charges in conductive objects to separate due to electrostatic induction. Negative charges *(blue)* are attracted and move to the surface of the object facing the external charge. Positive charges *(red)* are repelled and move to the surface facing away. These induced surface charges are exactly the right size and shape so their opposing electric field cancels the electric field of the external charge throughout the interior of the metal. Therefore, the electrostatic field everywhere inside a conductive object is zero, and the electrostatic potential is constant.

$$\vec{F} = q\vec{E}.$$

(See the Lorentz equation if the charge is not stationary.)

Consider a collection of N particles of charge Q_i, located at points $\vec{r}_i$ (called *source points*), the electric field at $\vec{r}$ (called the *field point*) is:

$$\vec{E}(\vec{r}) = \frac{1}{4\pi\varepsilon_0} \sum_{i=1}^{N} \frac{\widehat{\mathcal{R}}_i Q_i}{\left\|\mathcal{R}_i\right\|^2},$$

where $\vec{\mathcal{R}}_i = \vec{r} - \vec{r}_i$, is the displacement vector from a *source point* $\vec{r}_i$ to the *field point* $\vec{r}$, and $\hat{\mathcal{R}}_i = \vec{\mathcal{R}}_i / \|\vec{\mathcal{R}}_i\|$ is a unit vector that indicates the direction of the field. For a single point charge at the origin, the magnitude of this electric field is $E = k_e Q / \mathcal{R}^2$, and points away from that charge is positive. The fact that the force (and hence the field) can be calculated by summing over all the contributions due to individual source particles is an example of the superposition principle. The electric field produced by a distribution of charges is given by the volume charge density $\rho(\vec{r})$ and can be obtained by converting this sum into a triple integral:

Gauss's Law

Gauss's law states that " the total electric flux through any closed surface in free space of any shape drawn in an electric field is proportional to the total electric charge enclosed by the surface." Mathematically, Gauss's law takes the form of an integral equation:

$$\oint_S \vec{E} \cdot d\vec{A} = \frac{1}{\varepsilon_0} Q_{enclosed} = \int_V \frac{\rho}{\varepsilon_0} \cdot d^3 r,$$

where $d^3 r = dx\, dy\, dz$ is a volume element. If the charge is distributed over a surface or along a line, replace $\rho d^3 r$ by σdA or $\lambda d\ell$. The Divergence Theorem allows Gauss's Law to be written in differential form:

$$\vec{\nabla} \cdot \vec{E} = \frac{\rho}{\varepsilon_0}.$$

where $\vec{\nabla} \cdot$ is the divergence operator.

Poisson and Laplace Equations

The definition of electrostatic potential, combined with the differential form of Gauss's law (above), provides a relationship between the potential Φ and the charge density ρ:

$$\nabla^2 \phi = -\frac{\rho}{\varepsilon_0}.$$

This relationship is a form of Poisson's equation. In the absence of unpaired electric charge, the equation becomes Laplace's equation:

$$\nabla^2 \phi = 0,$$

Electrostatic Approximation

The validity of the electrostatic approximation rests on the assumption that the electric field is irrotational:

$$\vec{\nabla} \times \vec{E} = 0.$$

From Faraday's law, this assumption implies the absence or near-absence of time-varying magnetic fields:

$$\frac{\partial \vec{B}}{\partial t} = 0.$$

In other words, electrostatics does not require the absence of magnetic fields or electric currents. Rather, if magnetic fields or electric currents *do* exist, they must not change with time, or in the worst-case, they must change with time only *very slowly*. In some problems, both electrostatics and magnetostatics may be required for accurate predictions, but the coupling between the two can still be ignored. Electrostatics and magnetostatics can both be seen as Galinean limits for electromagnetism.

Electrostatic Potential

Because the electric field is irrotational, it is possible to express the electric field as the gradient of a scalar function, ϕ, called the electrostatic potential (also known as the voltage). An electric field, E, points from regions of high electric potential to regions of low electric potential, expressed mathematically as

$$\vec{E} = -\vec{\nabla}\phi.$$

The Gradient Theorem can be used to establish that the electrostatic potential is the amount of work per unit charge required to move a charge from point a to point b is the following line integral:

$$-\int_a^b \vec{E} \cdot \mathrm{d}\vec{\ell} = \phi(\vec{b}) - \phi(\vec{a}).$$

From these equations, we see that the electric potential is constant in any region for which the electric field vanishes (such as occurs inside a conducting object).

Electrostatic Energy

A single test particle's potential energy, U_E^{single}, can be calculated from a line integral of the work, $q_n \vec{E} \cdot \mathrm{d}\vec{\ell}$. We integrate from a point at infinity, and assume a collection of N particles of charge Q_n, are already situated at the points $\vec{r}_i$. This potential energy (in Joules) is:

$$U_E^{\text{single}} = q\phi(\vec{r}) = \frac{q}{4\pi\varepsilon_0}\sum_{i=1}^N \frac{Q_i}{\left\|\vec{R}_i\right\|}$$

where $\vec{R}_i = \vec{r} - \vec{r}_i$ is the distance of each charge Q_i from the test charge q, which situated at the point $\vec{r}$, and $\phi(\vec{r})$ is the electric potential that would be at $\vec{r}$ if the test charge were not present. If only two charges are present, the potential energy is $k_e Q_1 Q_2 / r$. The total electric potential energy due a collection of N charges is calculating by assembling these particles one at a time:

$$U_E^{\text{total}} = \frac{1}{4\pi\varepsilon_0}\sum_{j=1}^N Q_j \sum_{i=1}^{j-1} \frac{Q_i}{r_{ij}} = \frac{1}{2}\sum_{i=1}^N Q_i \phi_i,$$

where the following sum from, $j = 1$ to N, excludes $i = j$:

$$\phi_i = \frac{1}{4\pi\varepsilon_0}\sum_{j=1(j\neq i)}^N \frac{Q_j}{r_{ij}}.$$

This electric potential, ϕ_i is what would be measured at $\vec{r}_i$ if the charge Q_i were missing. This formula obviously excludes the (infinite) energy that would be required to assemble each point charge from a disperse cloud of charge. The sum over charges can be converted into an integral over charge density using the prescription $\sum(\cdots) \to \int(\cdots)\rho\mathrm{d}^3 r$:

$$U_E^{\text{total}} = \frac{1}{2}\int \rho(\vec{r})\phi(\vec{r})\mathrm{d}^3 r = \frac{\varepsilon_0}{2}\int |\mathbf{E}|^2 \mathrm{d}^3 r,,$$

This second expression for electrostatic energy uses the fact that the electric field is the negative gradient of the electric potential, as well as vector calculus identities in a way that resembles integration by parts. These two integrals for electric field energy seem to indicate two mutually exclusive formulas for electrostatic energy density, namely $\frac{1}{2}\rho\phi$ and $\frac{\varepsilon_0}{2}E^2$; they yield equal values for the total electrostatic energy only if both are integrated over all space.

Electrostatic Pressure

On a conductor, a surface charge will experience a force in the presence of an electric field. This force is the average of the discontinuous electric field at the surface charge. This average in terms of the field just outside the surface amounts to:

$$P = \frac{\varepsilon_0}{2} E^2 ,$$

This pressure tends to draw the conductor into the field, regardless of the sign of the surface charge.

Triboelectric Series

The triboelectric effect is a type of contact electrification in which certain materials become electrically charged when they are brought into contact with a different material and then separated. One of the materials acquires a positive charge, and the other acquires an equal negative charge. The polarity and strength of the charges produced differ according to the materials, surface roughness, temperature, strain, and other properties. Amber, for example, can acquire an electric charge by friction with a material like wool. This property, first recorded by Thales of Miletus, was the first electrical phenomenon investigated by humans. Other examples of materials that can acquire a significant charge when rubbed together include glass rubbed with silk, and hard rubber rubbed with fur.

Electrostatic Generators

The presence of surface charge imbalance means that the objects will exhibit attractive or repulsive forces. This surface charge imbalance, which yields static electricity, can be generated by touching two differing surfaces together and then separating them due to the phenomena of contact electrification and the triboelectric effect. Rubbing two nonconductive objects generates a great amount of static electricity. This is not just the result of friction; two nonconductive surfaces can become charged by just being placed one on top of the other. Since most surfaces have a rough texture, it takes longer to achieve charging through contact than through rubbing. Rubbing objects together increases amount of adhesive contact between the two surfaces. Usually insulators, e.g., substances that do not conduct electricity, are good at both generating, and holding, a surface charge. Some examples of these substances are rubber, plastic, glass, and pith. Conductive objects only rarely generate charge imbalance except, for example, when a metal surface is impacted by solid or liquid nonconductors. The charge that is transferred during contact electrification is stored on the surface of each object. Static electric generators, devices which produce very high voltage at very low current and used for classroom physics demonstrations, rely on this effect.

Note that the presence of electric current does not detract from the electrostatic forces nor from the sparking, from the corona discharge, or other phenomena. Both phenomena can exist simultaneously in the same system.

Charge Neutralization

Natural electrostatic phenomena are most familiar as an occasional annoyance in seasons of low humidity, but can be destructive and harmful in some situations (e.g. electronics manufacturing).

When working in direct contact with integrated circuit electronics (especially delicate MOSFETs), or in the presence of flammable gas, care must be taken to avoid accumulating and suddenly discharging a static charge.

Charge Induction

Charge induction occurs when a negatively charged object repels (the negatively charged) electrons from the surface of a second object. This creates a region in the second object that is more positively charged. An attractive force is then exerted between the objects. For example, when a balloon is rubbed, the balloon will stick to the wall as an attractive force is exerted by two oppositely charged surfaces (the surface of the wall gains an electric charge due to charge induction, as the free electrons at the surface of the wall are repelled by the negative balloon, creating a positive wall surface, which is subsequently attracted to the surface of the balloon). You can explore the effect with a simulation of the balloon and static electricity.

'Static' Electricity

Before the year 1832, when Michael Faraday published the results of his experiment on the identity of electricities, physicists thought "static electricity" was somehow different from other electrical charges. Michael Faraday proved that the electricity induced from the magnet, voltaic electricity produced by a battery, and static electricity are all the same.

Lightning over Oradea in Romania

Static electricity is usually caused when certain materials are rubbed against each other, like wool on plastic or the soles of shoes on carpet. The process causes electrons to be pulled from the surface of one material and relocated on the surface of the other material.

A static shock occurs when the surface of the second material, negatively charged with electrons, touches a positively charged conductor, or vice versa.

Static electricity is commonly used in xerography, air filters, and some automotive paints. Static electricity is a buildup of electric charges on two objects that have become separated from each other. Small electrical components can easily be damaged by static electricity. Component manufacturers use a number of antistatic devices to avoid this.

Static Electricity and Chemical Industry

When different materials are brought together and then separated, an accumulation of electric charge can occur which leaves one material positively charged while the other becomes negatively charged. The mild shock that you receive when touching a grounded object after walking on carpet is an example of excess electrical charge accumulating in your body from frictional charging between your shoes and the carpet. The resulting charge build-up upon your body can generate a strong electrical discharge. Although experimenting with static electricity may be fun, similar sparks create severe hazards in those industries dealing with flammable substances, where a small electrical spark may ignite explosive mixtures with devastating consequences.

A similar charging mechanism can occur within low conductivity fluids flowing through pipelines—a process called flow electrification. Fluids which have low electrical conductivity (below 50 picosiemens per meter), are called accumulators. Fluids having conductivities above 50 pS/m are called non-accumulators. In non-accumulators, charges recombine as fast as they are separated and hence electrostatic charge generation is not significant. In the petrochemical industry, 50 pS/m is the recommended minimum value of electrical conductivity for adequate removal of charge from a fluid.

An important concept for insulating fluids is the static relaxation time. This is similar to the time constant (tau) within an RC circuit. For insulating materials, it is the ratio of the static dielectric constant divided by the electrical conductivity of the material. For hydrocarbon fluids, this is sometimes approximated by dividing the number 18 by the electrical conductivity of the fluid. Thus a fluid that has an electrical conductivity of 1 pS/cm (100 pS/m) will have an estimated relaxation time of about 18 seconds. The excess charge within a fluid will be almost completely dissipated after 4 to 5 times the relaxation time, or 90 seconds for the fluid in the above example.

Charge generation increases at higher fluid velocities and larger pipe diameters, becoming quite significant in pipes 8 inches (200 mm) or larger. Static charge generation in these systems is best controlled by limiting fluid velocity. The British standard BS PD CLC/TR 50404:2003 (formerly BS-5958-Part 2) Code of Practice for Control of Undesirable Static Electricity prescribes velocity limits. Because of its large impact on dielectric constant, the recommended velocity for hydrocarbon fluids containing water should be limited to 1 m/s.

Bonding and earthing are the usual ways by which charge buildup can be prevented. For fluids with electrical conductivity below 10 pS/m, bonding and earthing are not adequate for charge dissipation, and anti-static additives may be required.

Applicable Standards

1. BS PD CLC/TR 50404:2003 Code of Practice for Control of Undesirable Static Electricity

2. NFPA 77 (2007) Recommended Practice on Static Electricity

3. API RP 2003 (1998) Protection Against Ignitions Arising Out of Static, Lightning, and Stray Currents

Electrostatic Induction in Commercial Applications

Electrostatic induction was used in the past to build high-voltage generators known as Influence machines. The main component that emerged in theses times is the capacitor. Electrostatic induction is also used for electromechanic precipitation or projection. In such technologies, charged particles of small sizes are collected or deposited intentionally on surfaces. Applications range from Electrostatic precipitator to Spray painting or Inkjet printing. Recently a new Wireless power Transfer Technology has been based on electrostatic induction between oscillating distant dipoles.

Piezoelectricity

Piezoelectricity is the electric charge that accumulates in certain solid materials (such as crystals, certain ceramics, and biological matter such as bone, DNA and various proteins) in response to applied mechanical stress. The word *piezoelectricity* means electricity resulting from pressure. Piezoelectricity was discovered in 1880 by French physicists Jacques and Pierre Curie.

The piezoelectric effect is understood as the linear electromechanical interaction between the mechanical and the electrical state in crystalline materials with no inversion symmetry. The piezoelectric effect is a reversible process in that materials exhibiting the direct piezoelectric effect (the internal generation of electrical charge resulting from an applied mechanical force) also exhibit the reverse piezoelectric effect (the internal generation of a mechanical strain resulting from an applied electrical field). For example, lead zirconate titanate crystals will generate measurable piezoelectricity when their static structure is deformed by about 0.1% of the original dimension. Conversely, those same crystals will change about 0.1% of their static dimension when an external electric field is applied to the material. The inverse piezoelectric effect is used in production of ultrasonic sound waves.

Piezoelectricity is found in useful applications, such as the production and detection of sound, generation of high voltages, electronic frequency generation, microbalances, to drive an ultrasonic nozzle, and ultrafine focusing of optical assemblies. It is also the basis of a number of scientific instrumental techniques with atomic resolution, the scanning probe microscopies, such as STM, AFM, MTA, SNOM, etc., and everyday uses, such as acting as the ignition source for cigarette lighters, and push-start propane barbecues, as well as the time reference source in quartz watches.

History

Discovery and Early Research

The pyroelectric effect, by which a material generates an electric potential in response to a temperature change, was studied by Carl Linnaeus and Franz Aepinus in the mid-18th century. Drawing on this knowledge, both René Just Haüy and Antoine César Becquerel posited a relationship between mechanical stress and electric charge; however, experiments by both proved inconclusive.

The first demonstration of the direct piezoelectric effect was in 1880 by the brothers Pierre Curie and Jacques Curie. They combined their knowledge of pyroelectricity with their understanding of the underlying crystal structures that gave rise to pyroelectricity to predict crystal behavior, and demonstrated the effect using crystals of tourmaline, quartz, topaz, cane sugar, and Rochelle salt (sodium potassium tartrate tetrahydrate). Quartz and Rochelle salt exhibited the most piezoelectricity.

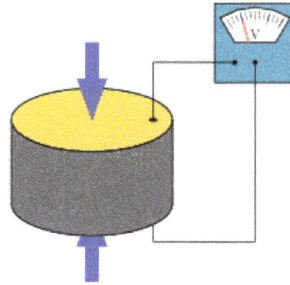

A piezoelectric disk generates a voltage when deformed (change in shape is greatly exaggerated)

The Curies, however, did not predict the converse piezoelectric effect. The converse effect was mathematically deduced from fundamental thermodynamic principles by Gabriel Lippmann in 1881. The Curies immediately confirmed the existence of the converse effect, and went on to obtain quantitative proof of the complete reversibility of electro-elasto-mechanical deformations in piezoelectric crystals.

For the next few decades, piezoelectricity remained something of a laboratory curiosity. More work was done to explore and define the crystal structures that exhibited piezoelectricity. This culminated in 1910 with the publication of Woldemar Voigt's *Lehrbuch der Kristallphysik* (*Textbook on Crystal Physics*), which described the 20 natural crystal classes capable of piezoelectricity, and rigorously defined the piezoelectric constants using tensor analysis.

World War I and Post-war

The first practical application for piezoelectric devices was sonar, first developed during World War I. In France in 1917, Paul Langevin and his coworkers developed an ultrasonic submarine detector. The detector consisted of a transducer, made of thin quartz crystals carefully glued between two steel plates, and a hydrophone to detect the returned echo. By emitting a high-frequency pulse from the transducer, and measuring the amount of time it takes to hear an echo from the sound waves bouncing off an object, one can calculate the distance to that object.

The use of piezoelectricity in sonar, and the success of that project, created intense development interest in piezoelectric devices. Over the next few decades, new piezoelectric materials and new applications for those materials were explored and developed.

Piezoelectric devices found homes in many fields. Ceramic phonograph cartridges simplified player design, were cheap and accurate, and made record players cheaper to maintain and easier to build. The development of the ultrasonic transducer allowed for easy measurement of viscosity and elasticity in fluids and solids, resulting in huge advances in materials research. Ultrasonic time-domain reflectometers (which send an ultrasonic pulse through a material and measure reflections from discontinuities) could find flaws inside cast metal and stone objects, improving structural safety.

World War II and Post-war

During World War II, independent research groups in the United States, Russia, and Japan discovered a new class of synthetic materials, called ferroelectrics, which exhibited piezoelectric constants many times higher than natural materials. This led to intense research to develop barium titanate and later lead zirconate titanate materials with specific properties for particular applications.

One significant example of the use of piezoelectric crystals was developed by Bell Telephone Laboratories. Following World War I, Frederick R. Lack, working in radio telephony in the engineering department, developed the "AT cut" crystal, a crystal that operated through a wide range of temperatures. Lack's crystal didn't need the heavy accessories previous crystal used, facilitating its use on aircraft. This development allowed Allied air forces to engage in coordinated mass attacks through the use of aviation radio.

Development of piezoelectric devices and materials in the United States was kept within the companies doing the development, mostly due to the wartime beginnings of the field, and in the interests of securing profitable patents. New materials were the first to be developed — quartz crystals were the first commercially exploited piezoelectric material, but scientists searched for higher-performance materials. Despite the advances in materials and the maturation of manufacturing processes, the United States market did not grow as quickly as Japan's did. Without many new applications, the growth of the United States' piezoelectric industry suffered.

In contrast, Japanese manufacturers shared their information, quickly overcoming technical and manufacturing challenges and creating new markets. In Japan, a temperature stable crystal cut was developed by Issac Koga. Japanese efforts in materials research created piezoceramic materials competitive to the U.S. materials but free of expensive patent restrictions. Major Japanese piezoelectric developments included new designs of piezoceramic filters for radios and televisions, piezo buzzers and audio transducers that can connect directly to electronic circuits, and the piezoelectric igniter, which generates sparks for small engine ignition systems (and gas-grill lighters) by compressing a ceramic disc. Ultrasonic transducers that transmit sound waves through air had existed for quite some time but first saw major commercial use in early television remote controls. These transducers now are mounted on several car models as an echolocation device, helping the driver determine the distance from the rear of the car to any objects that may be in its path.

Mechanism

The nature of the piezoelectric effect is closely related to the occurrence of electric dipole moments in solids. The latter may either be induced for ions on crystal lattice sites with asymmetric charge surroundings (as in $BaTiO_3$ and PZTs) or may directly be carried by molecular groups (as in cane sugar). The dipole density or polarization (dimensionality [Cm/m^3]) may easily be calculated for crystals by summing up the dipole moments per volume of the crystallographic unit cell. As every dipole is a vector, the dipole density P is a vector field. Dipoles near each other tend to be aligned in regions called Weiss domains. The domains are usually randomly oriented, but can be aligned using the process of *poling* (not the same as magnetic poling), a process by which a strong electric field is applied across the material, usually at elevated temperatures. Not all piezoelectric materials can be poled.

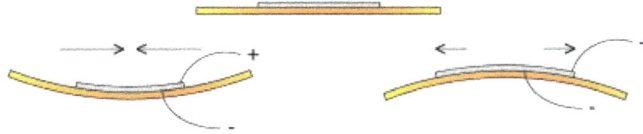

Piezoelectric plate used to convert audio signal to sound waves

Of decisive importance for the piezoelectric effect is the change of polarization P when applying a mechanical stress. This might either be caused by a reconfiguration of the dipole-inducing surrounding or by re-orientation of molecular dipole moments under the influence of the external stress. Piezoelectricity may then manifest in a variation of the polarization strength, its direction or both, with the details depending on: 1. the orientation of P within the crystal; 2. crystal symmetry; and 3. the applied mechanical stress. The change in P appears as a variation of surface charge density upon the crystal faces, i.e. as a variation of the electric field extending between the faces caused by a change in dipole density in the bulk. For example, a 1 cm³ cube of quartz with 2 kN (500 lbf) of correctly applied force can produce a voltage of 12500 V.

Piezoelectric materials also show the opposite effect, called the converse piezoelectric effect, where the application of an electrical field creates mechanical deformation in the crystal.

Mathematical Description

Linear piezoelectricity is the combined effect of

- The linear electrical behavior of the material:

$$\mathbf{D} = \varepsilon \mathbf{E} \quad \Rightarrow \quad D_i = \varepsilon_{ij} E_j$$

  where D is the electric charge density displacement (electric displacement), ε is permittivity (free-body dielectric constant), E is electric field strength, and $\nabla \cdot \mathbf{D} = 0, \nabla \times \mathbf{E} = \mathbf{0}$.

- Hooke's Law for linear elastic materials:

$$\mathbf{S} = \mathbf{s}\mathbf{T} \quad \Rightarrow \quad S_{ij} = s_{ijkl} T_{kl}$$

  where S is strain, s is compliance under short-circuit conditions, T is stress, and

$$\nabla \cdot \boldsymbol{T} = \boldsymbol{0}, \boldsymbol{S} = \frac{\nabla \mathbf{u} + \mathbf{u}\nabla}{2}.$$

These may be combined into so-called *coupled equations*, of which the strain-charge form is:

$$\mathbf{S} = \mathbf{s}\mathbf{T} + \eth^t \mathbf{E} \quad \Rightarrow \quad S_{ij} = s_{ijkl} T_{kl} + d_{kij} E_k$$

$$\mathbf{D} = \quad \mathbf{T} + \quad \Rightarrow \quad D_i = d_{ijk} T_{jk} + {}_{ij} E_j$$

In matrix form,

$$\{S\} = \left[s^E \right] \{T\} + [d^t]\{E\}$$

$$\{D\} = [d]\{T\} + \left[\varepsilon^T\right]\{E\},$$

where $[d]$ is the matrix for the direct piezoelectric effect and $[d^t]$ is the matrix for the converse piezoelectric effect. The superscript E indicates a zero, or constant, electric field; the superscript T indicates a zero, or constant, stress field; and the superscript t stands for transposition of a matrix.

Notice that the third order tensor maps vectors into symmetric matrices. There are no non-trivial rotation-invariant tensors that have this property, which is why there are no isotropic piezoelectric materials.

The strain-charge for a material of the 4mm (C_{4v}) crystal class (such as a poled piezoelectric ceramic such as tetragonal PZT or BaTiO$_3$) as well as the 6mm crystal class may also be written as (ANSI IEEE 176):

$$\begin{bmatrix} S_1 \\ S_2 \\ S_3 \\ S_4 \\ S_5 \\ S_6 \end{bmatrix} = \begin{bmatrix} s_{11}^E & s_{12}^E & s_{13}^E & 0 & 0 & 0 \\ s_{21}^E & s_{22}^E & s_{23}^E & 0 & 0 & 0 \\ s_{31}^E & s_{32}^E & s_{33}^E & 0 & 0 & 0 \\ 0 & 0 & 0 & s_{44}^E & 0 & 0 \\ 0 & 0 & 0 & 0 & s_{55}^E & 0 \\ 0 & 0 & 0 & 0 & 0 & s_{66}^E = 2\left(s_{11}^E - s_{12}^E\right) \end{bmatrix} \begin{bmatrix} T_1 \\ T_2 \\ T_3 \\ T_4 \\ T_5 \\ T_6 \end{bmatrix} + \begin{bmatrix} 0 & 0 & d_{31} \\ 0 & 0 & d_{32} \\ 0 & 0 & d_{33} \\ 0 & d_{24} & 0 \\ d_{15} & 0 & 0 \\ 0 & 0 & 0 \end{bmatrix} \begin{bmatrix} E_1 \\ E_2 \\ E_3 \end{bmatrix}$$

$$\begin{bmatrix} D_1 \\ D_2 \\ D_3 \end{bmatrix} = \begin{bmatrix} 0 & 0 & 0 & 0 & d_{15} & 0 \\ 0 & 0 & 0 & d_{24} & 0 & 0 \\ d_{31} & d_{32} & d_{33} & 0 & 0 & 0 \end{bmatrix} \begin{bmatrix} T_1 \\ T_2 \\ T_3 \\ T_4 \\ T_5 \\ T_6 \end{bmatrix} + \begin{bmatrix} \varepsilon_{11} & 0 & 0 \\ 0 & \varepsilon_{22} & 0 \\ 0 & 0 & \varepsilon_{33} \end{bmatrix} \begin{bmatrix} E_1 \\ E_2 \\ E_3 \end{bmatrix}$$

where the first equation represents the relationship for the converse piezoelectric effect and the latter for the direct piezoelectric effect.

Although the above equations are the most used form in literature, some comments about the notation are necessary. Generally, D and E are vectors, that is, Cartesian tensors of rank 1; and permittivity ε is a Cartesian tensor of rank 2. Strain and stress are, in principle, also rank-2 tensors. But conventionally, because strain and stress are all symmetric tensors, the subscript of strain and stress can be relabeled in the following fashion: $11 \rightarrow 1$; $22 \rightarrow 2$; $33 \rightarrow 3$; $23 \rightarrow 4$; $13 \rightarrow 5$; $12 \rightarrow 6$. (Different conventions may be used by different authors in literature. For example, some use $12 \rightarrow 4$; $23 \rightarrow 5$; $31 \rightarrow 6$ instead.) That is why S and T appear to have the "vector form" of six components. Consequently, s appears to be a 6-by-6 matrix instead of a rank-4 tensor. Such a relabeled notation is often called Voigt notation. Whether the shear strain components S_4, S_5, S_6 are tensor components or engineering strains is another question. In the equation above, they must be engineering strains for the 6,6 coefficient of the compliance matrix to be written as shown, i.e., $2(sE11 - sE12)$. Engineering shear strains are double the value of the corresponding tensor shear, such as $S_6 = 2S_{12}$ and so on. This also means that $s_{66} = 1/G_{12}$, where G_{12} is the shear modulus.

In total, there are four piezoelectric coefficients, d_{ij}, e_{ij}, g_{ij}, and h_{ij} defined as follows:

$$d_{ij} = \left(\frac{\partial D_i}{\partial T_j}\right)^E = \left(\frac{\partial S_j}{\partial E_i}\right)^T$$

$$e_{ij} = \left(\frac{\partial D_i}{\partial S_j}\right)^E = -\left(\frac{\partial T_j}{\partial E_i}\right)^S$$

$$g_{ij} = -\left(\frac{\partial E_i}{\partial T_j}\right)^D = \left(\frac{\partial S_j}{\partial D_i}\right)^T$$

$$h_{ij} = -\left(\frac{\partial E_i}{\partial S_j}\right)^D = -\left(\frac{\partial T_j}{\partial D_i}\right)^S$$

where the first set of four terms corresponds to the direct piezoelectric effect and the second set of four terms corresponds to the converse piezoelectric effect. For those piezoelectric crystals for which the polarization is of the crystal-field induced type, a formalism has been worked out that allows for the calculation of piezoelectrical coefficients d_{ij} from electrostatic lattice constants or higher-order Madelung constants.

Crystal Classes

Of the 32 crystal classes, 21 are non-centrosymmetric (not having a centre of symmetry), and of these, 20 exhibit direct piezoelectricity (the 21st is the cubic class 432). Ten of these represent the polar crystal classes, which show a spontaneous polarization without mechanical stress due to a non-vanishing electric dipole moment associated with their unit cell, and which exhibit pyroelectricity. If the dipole moment can be reversed by the application of an electric field, the material is said to be ferroelectric.

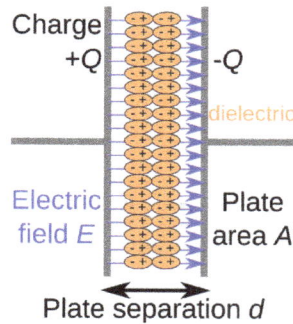

Any spatially separated charge will result in an electric field, and therefore an electric potential. Shown here is a standard dielectric in a capacitor. In a piezoelectric device, mechanical stress, instead of an externally applied voltage, causes the charge separation in the individual atoms of the material.

- Polar crystal classes: 1, 2, m, mm2, 4, 4mm, 3, 3m, 6, 6mm.

- Piezoelectric crystal classes: 1, 2, m, 222, mm2, 4, 4, 422, 4mm, 42m, 3, 32, 3m, 6, 6, 622, 6mm, 62m, 23, 43m.

For polar crystals, for which $P \neq 0$ holds without applying a mechanical load, the piezoelectric effect manifests itself by changing the magnitude or the direction of P or both.

For the nonpolar but piezoelectric crystals, on the other hand, a polarization P different from zero is only elicited by applying a mechanical load. For them the stress can be imagined to transform the material from a nonpolar crystal class ($P = 0$) to a polar one, having $P \neq 0$.

Materials

Many materials, both natural and synthetic, exhibit piezoelectricity:

Naturally Occurring Crystals

- Quartz

- Berlinite ($AlPO_4$), a rare phosphate mineral that is structurally identical to quartz

- Sucrose (table sugar)

- Rochelle salt

- Topaz

- Tourmaline-group minerals

- Lead titanate ($PbTiO_3$). Although it occurs in nature as mineral macedonite, it is synthesized for research and applications.

The action of piezoelectricity in Topaz can probably be attributed to ordering of the (F,OH) in its lattice, which is otherwise centrosymmetric: orthorhombic bipyramidal (mmm). Topaz has anomalous optical properties which are attributed to such ordering.

Bone

Dry bone exhibits some piezoelectric properties. Studies of Fukada *et al.* showed that these are not due to the apatite crystals, which are centrosymmetric, thus non-piezoelectric, but due to collagen. Collagen exhibits the polar uniaxial orientation of molecular dipoles in its structure and can be considered as bioelectret, a sort of dielectric material exhibiting quasipermanent space charge and dipolar charge. Potentials are thought to occur when a number of collagen molecules are stressed in the same way displacing significant numbers of the charge carriers from the inside to the surface of the specimen. Piezoelectricity of single individual collagen fibrils was measured using piezoresponse force microscopy, and it was shown that collagen fibrils behave predominantly as shear piezoelectric materials.

The piezoelectric effect is generally thought to act as a biological force sensor. This effect was exploited by research conducted at the University of Pennsylvania in the late 1970s and early 1980s, which established that sustained application of electrical potential could stimulate both resorption and growth (depending on the polarity) of bone in-vivo. Further studies in the 1990s provided the mathematical equation to confirm long bone wave propagation as to that of hexagonal (Class 6) crystals.

Other Natural Materials

Biological materials exhibiting piezoelectric properties include:

- Tendon

- Silk

- Wood due to piezoelectric texture

- Enamel

- Dentin

- DNA

- Viral proteins, including those from bacteriophage. One study has found that thin films of M13 bacteriophage can be used to construct a piezoelectric generator sufficient to operate a liquid crystal display.

Synthetic Crystals

- Langasite ($La_3Ga_5SiO_{14}$), a quartz-analogous crystal

- Gallium orthophosphate ($GaPO_4$), a quartz-analogous crystal

- Lithium niobate ($LiNbO_3$)

- Lithium tantalate ($LiTaO_3$)

Synthetic Ceramics

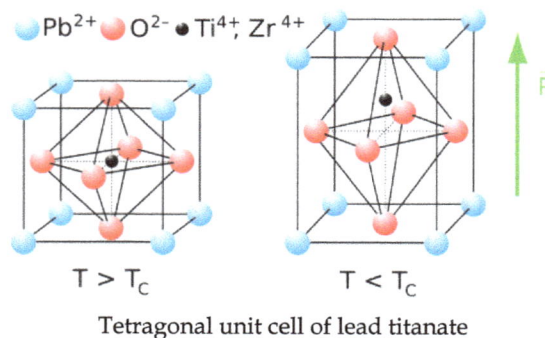

Tetragonal unit cell of lead titanate

Ceramics with randomly oriented grains must be ferroelectric to exhibit piezoelectricity. The macroscopic piezoelectricity is possible in textured polycrystalline non-ferroelectric piezoelectric materials, such as AlN and ZnO. The family of ceramics with perovskite, tungsten-bronze and related structures exhibits piezoelectricity:

- Barium titanate ($BaTiO_3$)—Barium titanate was the first piezoelectric ceramic discovered.

- Lead zirconate titanate ($Pb[Zr_{xTi1-x}]O_3$ with $0 \leq x \leq 1$)—more commonly known as PZT, lead zirconate titanate is the most common piezoelectric ceramic in use today.

- Potassium niobate ($KNbO_3$)

- Sodium tungstate (Na_2WO_3)

- $Ba_2NaNb_5O_5$

- $Pb_2KNb_5O_{15}$

- Zinc oxide (ZnO)–Wurtzite structure. While single crystals of ZnO are piezoelectric and pyroelectric, polycrystalline (ceramic) ZnO with randomly oriented grains exhibits neither piezoelectric nor pyroelectric effect. Not being ferroelectric, polycrystalline ZnO cannot be poled like barium titanate or PZT. Ceramics and polycrystalline thin films of ZnO may exhibit macroscopic piezoelectricity and pyroelectricity only if they are textured (grains are preferentially oriented), such that the piezoelectric and pyroelectric responses of all individual grains do not cancel. This is readily accomplished in polycrystalline thin films.

Lead-free Piezoceramics

More recently, there is growing concern regarding the toxicity in lead-containing devices driven by the result of restriction of hazardous substances directive regulations. To address this concern, there has been a resurgence in the compositional development of lead-free piezoelectric materials.

- Sodium potassium niobate (($K,Na)NbO_3$). This material is also known as NKN. In 2004, a group of Japanese researchers led by Yasuyoshi Saito discovered a sodium potassium niobate composition with properties close to those of PZT, including a high T_C. Certain compositions of this material have been shown to retain a high mechanical quality factor ($Q_m \approx 900$) with increasing vibration levels, whereas the mechanical quality factor of hard PZT degrades in such conditions. This fact makes NKN a promising replacement for high power resonance applications, such as piezoelectric transformers.

- Bismuth ferrite ($BiFeO_3$) is also a promising candidate for the replacement of lead-based ceramics.

- Sodium niobate $NaNbO_3$

- Bismuth titanate $Bi_4Ti_3O_{12}$

- Sodium bismuth titanate $NaBi(TiO_3)_2$

So far, neither the environmental effect nor the stability of supplying these substances have been measured.

III–V and II–VI Semiconductors

A piezoelectric potential can be created in any bulk or nanostructured semiconductor crystal having non central symmetry, such as the Group III–V and II–VI materials, due to polarization of ions under applied stress and strain. This property is common to both the zincblende and wurtzite crystal structures. To first order, there is only one independent piezoelectric coefficient in zinc-blende, called e_{14}, coupled to shear components of the strain. In wurtzite, there are instead three independent piezoelectric coefficients: e_{31}, e_{33} and e_{15}. The semiconductors where the strongest piezoelectricity is observed are those commonly found in the wurtzite structure, i.e. GaN, InN, AlN and ZnO. ZnO is the most used material in the recent field of piezotronics.

Since 2006, there have also been a number of reports of strong non linear piezoelectric effects in polar semiconductors. Such effects are generally recognized to be at least important if not of the same order of magnitude as the first order approximation.

Polymers

- Polyvinylidene fluoride (PVDF): PVDF exhibits piezoelectricity several times greater than quartz. Unlike ceramics, where the crystal structure of the material creates the piezoelectric effect, in polymers the intertwined long-chain molecules attract and repel each other when an electric field is applied.

Organic Nanostructures

A strong shear piezoelectric activity was observed in self-assembled diphenylalanine peptide nanotubes (PNTs), indicating electric polarization directed along the tube axis. Comparison with LiNbO$_3$ and lateral signal calibration yields sufficiently high effective piezoelectric coefficient values of at least 60 pm/V (shear response for tubes of ≈200 nm in diameter). PNTs demonstrate linear deformation without irreversible degradation in a broad range of driving voltages.

Application

Currently, industrial and manufacturing is the largest application market for piezoelectric devices, followed by the automotive industry. Strong demand also comes from medical instruments as well as information and telecommunications. The global demand for piezoelectric devices was valued at approximately US$14.8 billion in 2010. The largest material group for piezoelectric devices is piezocrystal, and piezopolymer is experiencing the fastest growth due to its low weight and small size.

Piezoelectric crystals are now used in numerous ways:

High Voltage and Power Sources

Direct piezoelectricity of some substances, like quartz, can generate potential differences of thousands of volts.

- The best-known application is the electric cigarette lighter: pressing the button causes a spring-loaded hammer to hit a piezoelectric crystal, producing a sufficiently high-voltage electric current that flows across a small spark gap, thus heating and igniting the gas. The portable sparkers used to ignite gas stoves work the same way, and many types of gas burners now have built-in piezo-based ignition systems.

- A similar idea is being researched by DARPA in the United States in a project called *Energy Harvesting*, which includes an attempt to power battlefield equipment by piezoelectric generators embedded in soldiers' boots. However, these energy harvesting sources by association affect the body. DARPA's effort to harness 1–2 watts from continuous shoe impact while walking were abandoned due to the impracticality and the discomfort from the additional energy expended by a person wearing the shoes. Other energy harvesting ideas include harvesting the energy from human movements in train stations or other public places and converting a dance floor to generate electricity. Vibrations from industrial machinery can also be harvested by piezoelectric materials to charge batteries for backup supplies or to power low-power microprocessors and wireless radios.

• A piezoelectric transformer is a type of AC voltage multiplier. Unlike a conventional transformer, which uses magnetic coupling between input and output, the piezoelectric transformer uses acoustic coupling. An input voltage is applied across a short length of a bar of piezoceramic material such as PZT, creating an alternating stress in the bar by the inverse piezoelectric effect and causing the whole bar to vibrate. The vibration frequency is chosen to be the resonant frequency of the block, typically in the 100 kilohertz to 1 megahertz range. A higher output voltage is then generated across another section of the bar by the piezoelectric effect. Step-up ratios of more than 1,000:1 have been demonstrated. An extra feature of this transformer is that, by operating it above its resonant frequency, it can be made to appear as an inductive load, which is useful in circuits that require a controlled soft start. These devices can be used in DC–AC inverters to drive cold cathode fluorescent lamps. Piezo transformers are some of the most compact high voltage sources.

Sensors

The principle of operation of a piezoelectric sensor is that a physical dimension, transformed into a force, acts on two opposing faces of the sensing element. Depending on the design of a sensor, different "modes" to load the piezoelectric element can be used: longitudinal, transversal and shear.

Many rocket-propelled grenades used a piezoelectric fuse. For example: RPG-7.

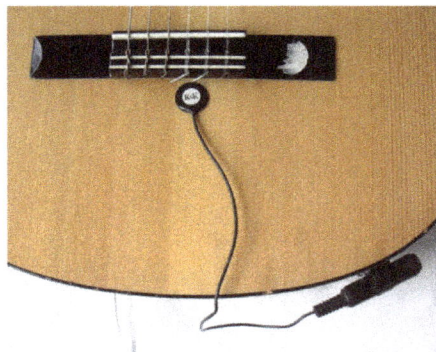

Piezoelectric disk used as a guitar pickup

Detection of pressure variations in the form of sound is the most common sensor application, e.g. piezoelectric microphones (sound waves bend the piezoelectric material, creating a changing voltage) and piezoelectric pickups for acoustic-electric guitars. A piezo sensor attached to the body of an instrument is known as a contact microphone.

Piezoelectric sensors especially are used with high frequency sound in ultrasonic transducers for medical imaging and also industrial nondestructive testing (NDT).

For many sensing techniques, the sensor can act as both a sensor and an actuator – often the term *transducer* is preferred when the device acts in this dual capacity, but most piezo devices have this property of reversibility whether it is used or not. Ultrasonic transducers, for example, can inject ultrasound waves into the body, receive the returned wave, and convert it to an electrical signal (a voltage). Most medical ultrasound transducers are piezoelectric.

In addition to those mentioned above, various sensor applications include:

- Piezoelectric elements are also used in the detection and generation of sonar waves.

- Piezoelectric materials are used in single-axis and dual-axis tilt sensing.

- Power monitoring in high power applications (e.g. medical treatment, sonochemistry and industrial processing).

- Piezoelectric microbalances are used as very sensitive chemical and biological sensors.

- Piezos are sometimes used in strain gauges.

- A piezoelectric transducer was used in the penetrometer instrument on the Huygens Probe.

- Piezoelectric transducers are used in electronic drum pads to detect the impact of the drummer's sticks, and to detect muscle movements in medical acceleromyography.

- Automotive engine management systems use piezoelectric transducers to detect Engine knock (Knock Sensor, KS), also known as detonation, at certain hertz frequencies. A piezoelectric transducer is also used in fuel injection systems to measure manifold absolute pressure (MAP sensor) to determine engine load, and ultimately the fuel injectors milliseconds of on time.

- Ultrasonic piezo sensors are used in the detection of acoustic emissions in acoustic emission testing.

Actuators

Metal disk with piezoelectric disk attached, used in a buzzer.

As very high electric fields correspond to only tiny changes in the width of the crystal, this width can be changed with better-than-μm precision, making piezo crystals the most important tool for positioning objects with extreme accuracy — thus their use in actuators. Multilayer ceramics, using layers thinner than 100 μm, allow reaching high electric fields with voltage lower than 150 V. These ceramics are used within two kinds of actuators: direct piezo actuators and Amplified piezoelectric actuators. While direct actuator's stroke is generally lower than 100 μm, amplified piezo actuators can reach millimeter strokes.

- Loudspeakers: Voltage is converted to mechanical movement of a metallic diaphragm.

- Piezoelectric motors: Piezoelectric elements apply a directional force to an axle, causing it to rotate. Due to the extremely small distances involved, the piezo motor is viewed as a high-precision replacement for the stepper motor.

- Piezoelectric elements can be used in laser mirror alignment, where their ability to move a large mass (the mirror mount) over microscopic distances is exploited to electronically align some laser mirrors. By precisely controlling the distance between mirrors, the laser electronics can accurately maintain optical conditions inside the laser cavity to optimize the beam output.

- A related application is the acousto-optic modulator, a device that scatters light off sound-waves in a crystal, generated by piezoelectric elements. This is useful for fine-tuning a laser's frequency.

- Atomic force microscopes and scanning tunneling microscopes employ converse piezoelectricity to keep the sensing needle close to the specimen.

- Inkjet printers: On many inkjet printers, piezoelectric crystals are used to drive the ejection of ink from the inkjet print head towards the paper.

- Diesel engines: High-performance common rail diesel engines use piezoelectric fuel injectors, first developed by Robert Bosch GmbH, instead of the more common solenoid valve devices.

- Active vibration control using amplified actuators.

- X-ray shutters.

- XY stages for micro scanning used in infrared cameras.

- Moving the patient precisely inside active CT and MRI scanners where the strong radiation or magnetism precludes electric motors.

- Crystal earpieces are sometimes used in old or low power radios.

- High-intensity focused ultrasound for localized heating or creating a localized cavitation can be achieved, for example, in patient's body or in an industrial chemical process.

Frequency Standard

The piezoelectrical properties of quartz are useful as a standard of frequency.

- Quartz clocks employ a crystal oscillator made from a quartz crystal that uses a combination of both direct and converse piezoelectricity to generate a regularly timed series of electrical pulses that is used to mark time. The quartz crystal (like any elastic material) has a precisely defined natural frequency (caused by its shape and size) at which it prefers to oscillate, and this is used to stabilize the frequency of a periodic voltage applied to the crystal.

- The same principle is critical in all radio transmitters and receivers, and in computers where it creates a clock pulse. Both of these usually use a frequency multiplier to reach gigahertz ranges.

Piezoelectric Motors

Types of piezoelectric motor include:

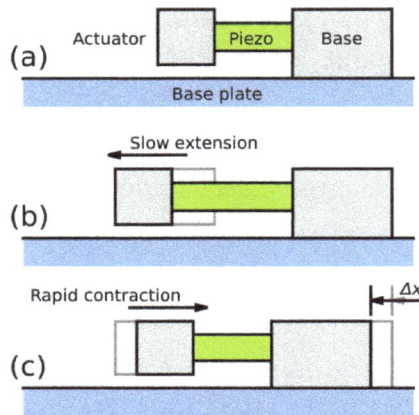

A slip-stick actuator.

- The traveling-wave motor used for auto-focus in reflex cameras

- Inchworm motors for linear motion

- Rectangular four-quadrant motors with high power density (2.5 W/cm^3) and speed ranging from 10 nm/s to 800 mm/s.

- Stepping piezo motor, using stick-slip effect.

Aside from the stepping stick-slip motor, all these motors work on the same principle. Driven by dual orthogonal vibration modes with a phase difference of 90°, the contact point between two surfaces vibrates in an elliptical path, producing a frictional force between the surfaces. Usually, one surface is fixed, causing the other to move. In most piezoelectric motors, the piezoelectric crystal is excited by a sine wave signal at the resonant frequency of the motor. Using the resonance effect, a much lower voltage can be used to produce a high vibration amplitude.

A stick-slip motor works using the inertia of a mass and the friction of a clamp. Such motors can be very small. Some are used for camera sensor displacement, thus allowing an anti-shake function.

Reduction of Vibrations and Noise

Different teams of researchers have been investigating ways to reduce vibrations in materials by attaching piezo elements to the material. When the material is bent by a vibration in one direction, the vibration-reduction system responds to the bend and sends electric power to the piezo element to bend in the other direction. Future applications of this technology are expected in cars and houses to reduce noise. Further applications to flexible structures, such as shells and plates, have also been studied for nearly three decades.

In a demonstration at the Material Vision Fair in Frankfurt in November 2005, a team from TU Darmstadt in Germany showed several panels that were hit with a rubber mallet, and the panel with the piezo element immediately stopped swinging.

Piezoelectric ceramic fiber technology is being used as an electronic damping system on some HEAD tennis rackets.

Infertility Treatment

In people with previous total fertilization failure, piezoelectric activation of oocytes together with intracytoplasmic sperm injection (ICSI) seems to improve fertilization outcomes.

Surgery

A recent application of piezoelectric ultrasound sources is piezoelectric surgery, also known as piezosurgery. Piezosurgery is a minimally invasive technique that aims to cut a target tissue with little damage to neighboring tissues. For example, Hoigne *et al.* reported its use in hand surgery for the cutting of bone, using frequencies in the range 25–29 kHz, causing microvibrations of 60–210 μm. It has the ability to cut mineralized tissue without cutting neurovascular tissue and other soft tissue, thereby maintaining a blood-free operating area, better visibility and greater precision.

Potential Applications

In 2015, Cambridge University researchers working in conjunction with researchers from the National Physical Laboratory and Cambridge-based dielectric antenna company Antenova Ltd, using thin films of piezoelectric materials found that at a certain frequency, these materials become not only efficient resonators, but efficient radiators as well, meaning that they can potentially be used as antennas. The researchers found that by subjecting the piezoelectric thin films to an asymmetric excitation, the symmetry of the system is similarly broken, resulting in a corresponding symmetry breaking of the electric field, and the generation of electromagnetic radiation.

In recent years, several attempts at the macro-scale application of the piezoelectric technology have emerged to harvest kinetic energy from walking pedestrians. The piezoelectric floors have been trialed since the beginning of 2007 in two Japanese train stations, Tokyo and Shibuya stations. The electricity generated from the foot traffic is used to provide all the electricity needed to run the automatic ticket gates and electronic display systems. In London, a famous nightclub exploited the piezoelectric technology in its dance floor. Parts of the lighting and sound systems in the club can be powered by the energy harvesting tiles. However, the piezoelectric tile deployed on the ground usually harvests energy from low frequency strikes provided by the foot traffic. This working condition may eventually lead to low power generation efficiency.

In this case, locating high traffic areas is critical for optimization of the energy harvesting efficiency, as well as the orientation of the tile pavement significantly affects the total amount of the harvested energy. A density flow evaluation is recommended to qualitatively evaluate the piezoelectric power harvesting potential of the considered area based on the number of pedestrian crossings per unit time. In X. Li's study, the potential application of a commercial piezoelectric energy harvester in a central hub building at Macquarie University in Sydney, Australia is examined and discussed. Optimization of the piezoelectric tile deployment is presented according to the frequency of pedestrian mobility and a model is developed where 3.1% of the total floor area with the highest pedestrian mobility is paved with piezoelectric tiles. The modelling results indicate that the total annual energy harvesting potential for the proposed optimized tile pavement model is estimated at 1.1 MW h/year, which would be sufficient to meet close to 0.5% of the annual energy needs of the building. In Israel, there is a company which has installed piezoelectric materials under a busy highway. The energy generated is adequate and powers street lights, billboards and signs.

Tyre company Goodyear has plans to develop an electricity generating tyre which has piezoelectric material lined inside it. As the tyre moves, it deforms and thus electricity is generated.

Photovoltaics

The efficiency of a hybrid photovoltaic cell that contains piezoelectric materials can be increased simply by placing it near a source of ambient noise or vibration. The effect was demonstrated with organic cells using zinc oxide nanotubes. The electricity generated by the piezoelectric effect itself is a negligible percentage of the overall output. Sound levels as low as 75 decibels improved efficiency by up to 50%. Efficiency peaked at 10 kHz, the resonant frequency of the nanotubes. The electrical field set up by the vibrating nanotubes interacts with electrons migrating from the organic polymer layer. This process decreases the likelihood of recombination, in which electrons are energized but settle back into a hole instead of migrating to the electron-accepting ZnO layer.

Sound Reinforcement System

A sound reinforcement system is the combination of microphones, signal processors, amplifiers, and loudspeakers that makes live or pre-recorded sounds louder and may also distribute those sounds to a larger or more distant audience. In some situations, a sound reinforcement system is also used to enhance or alter the sound of the sources on the stage, typically by using electronic effects, as opposed to simply amplifying the sources unaltered.

A sound reinforcement system for a rock concert in a stadium may be very complex, including hundreds of microphones, complex live sound mixing and signal processing systems, tens of thousands of watts of amplifier power, and multiple loudspeaker arrays, all overseen by a team of audio engineers and technicians. On the other hand, a sound reinforcement system can be as simple as a small public address (PA) system, consisting of, for example, a single microphone connected to an amplified loudspeaker for a singer-guitarist playing in a coffeehouse. In both cases, these systems *reinforce* sound to make it louder or distribute it to a wider audience.

Large outdoor concerts use complex and powerful sound reinforcement systems.

Some audio engineers and others in the professional audio industry disagree over whether these audio systems should be called sound reinforcement (SR) systems or PA systems. Distinguishing between the two terms by technology and capability is common, while others distinguish by intended use (e.g., SR systems are for live event support and PA systems are for reproduction of speech and recorded music in buildings and institutions). In some regions or markets, the distinction between the two terms is important, though the terms are considered interchangeable in many professional circles.

Basic Concept

A typical sound reinforcement system consists of; input transducers (e.g., microphones), which convert sound energy such as a person singing or talking or the sound of an instrument into an electric signal, signal processors which alter the signal characteristics (e.g., equalizers that adjust the bass and treble, compressors that reduce signal peaks, etc.), amplifiers, which add power to the signal so that the signal can drive a loudspeaker without otherwise changing its content, and output transducers (e.g., loudspeakers in speaker cabinets), which convert the signal back into sound energy (the sound heard by the audience and the performers). These primary parts involve varying amounts of individual components to achieve the desired goal of reinforcing and clarifying the sound to the audience, performers, or other individuals.

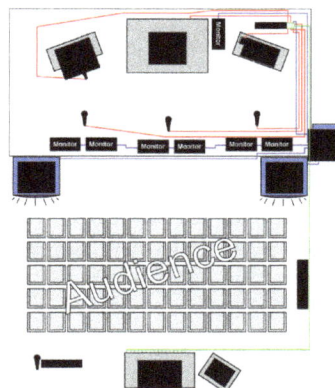

A basic sound reinforcement system that would be used in a small music venue. The main loudspeakers for the audience are to the left and right of the stage. A row of monitor speakers pointing towards the onstage performers helps them hear their singing and playing. The audio engineer sits at the back of the room, operating the mixing console which shapes the sound and volume of all of the voices and instruments.

Signal Path

Sound reinforcement in a large format system typically involves a signal path that starts with the signal inputs, which may be instrument pickups (on an electric guitar or electric bass) or a microphone (transducer) that a vocalist is singing into or a microphone placed in front of an instrument or guitar amplifier. These signal inputs are plugged into the input jacks of a thick multicore cable (often called a "snake"). The snake then routes the signals of all of the inputs to either one or two mixing consoles. In a coffeehouse or small nightclub, the snake may be only routed to a single mixing console, which an audio engineer will use to adjust the sound and volume of the onstage vocals and instruments that the audience hears through the main speakers and adjust the volume of the monitor speakers that are aimed at the performers.

Mid- to large-size performing venues typically route the onstage signals to two mixing consoles: the "Front of the House" (FOH) Main mix (which is sent to the speakers facing the audience), and the "Monitor mix", which is often a second mixer at the side of the stage. In these cases, at least two audio engineers are required; one to do the "FOH" Main mix and one to do the "Monitor mix".

Once the signal is at a channel on the mixing console, this signal can be equalized (e.g., by adjusting the bass or treble of the sound), compressed (to avoid unwanted signal peaks), or panned (that is sent to the left or right speakers) before being routed to an output bus. The signal may also be routed into an external effects processor, such as a reverb effect, which outputs a *wet* (effected) version of the signal, which is typically mixed in varying amounts with the *dry* (effect-free) signal.

The signal is then routed to a "bus", also known as a "mix group", "subgroup" or simply "group". A group of signals may be routed through an additional bus before being sent to the main bus to allow the engineer to control the levels of several related signals at once. For example, all of the different microphones for a drum set might be sent to their own bus so that the volume of the entire drum set sound can be controlled with a single fader or a pair of faders. A bus can often be processed just like an individual input channel, allowing the engineer to process a whole group of signals at once. The signal is then typically routed with everything else to the stereo master faders on the console. Mixing consoles also have additional "sends", also referred to as *auxes* or *aux sends* (an abbreviation for "auxiliary send"), on each input channel so that a different mix can be created and sent elsewhere. One usage for aux sends is to create a mix of the vocal and instrument signals for the monitor mix (this is what the onstage singers and musicians hear from their monitor speakers or in-ear monitors). Another use of an aux send is to select varying amounts of certain channels (via the aux send knobs on each channel), and then route these signals to an effects processor. A common example of the second use of aux sends is to send all of the vocal signals from a rock band through a reverb effect. While reverb is usually added to vocals in the main mix, it is not usually added to electric bass and other rhythm section instruments.

The next step in the signal path generally depends on the size of the system in place. In smaller systems, the main outputs are often sent to an additional equalizer, or directly to a power amplifier, with one or more loudspeakers (typically two, one on each side of the stage in smaller venues) then connected to that amplifier. In large-format systems, the signal is typically first routed through an equalizer then to a crossover. A crossover splits the signal into multiple frequency bands with each band being sent to separate amplifiers and speaker enclosures for low, middle, and high-frequency signals. Low-frequency signals are sent to amplifiers and then to subwoofers, and middle and high-frequency sounds are typically sent to amplifiers which power full-range speaker cabinets. Using a crossover to separate the sound into low, middle and high frequencies can lead to a "cleaner", clearer sound than routing all of the frequencies through a single full-range speaker system.

System Components

Input Transducers

Many types of input transducers can be found in a sound reinforcement system, with microphones being the most commonly used input device. Microphones can be classified according to their method of transduction, pickup (or polar) pattern or their functional application. Most microphones used in sound reinforcement are either dynamic or condenser microphones. One type of directional mics, called cardioid mics, are widely used in live sound, because they reduce pickup from the side and rear, helping to avoid unwanted feedback "howls" from the monitors.

Audio engineers use a range of microphones for different live sound applications.

Microphones used for sound reinforcement are positioned and mounted in many ways, including base-weighted upright stands, podium mounts, tie-clips, instrument mounts, and headset mounts. Microphones on stands are also placed in front of instruments and guitar amplifiers to pick up the sound. Headset mounted and tie-clip mounted microphones are often used with wireless transmission to allow performers or speakers to move freely. Early adopters of headset mounted microphones technology included country singer Garth Brooks, Kate Bush, and Madonna.

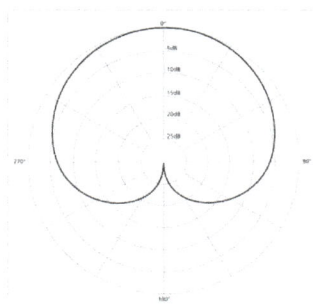

Cardioid mics are widely used in live sound, because their "apple-shaped" pickup pattern rejects sounds from the sides and rear of the mic, making it more resistant to unwanted feedback "howls".

There are many other types of input transducers which may be used occasionally, including magnetic pickups used in electric guitars and electric basses, which are plugged first into a DI box and then into the snake, contact microphones used on stringed instruments, and pianos and phonograph pickups (cartridges) used in record players. As well, some electronic instruments such as synthesizers can have their output signal routed to the mixing console.

A Yamaha PM4000 and a Midas Heritage 3000 mixing console at the Front of House position at an outdoor concert.

Wireless technology has become popular in sound reinforcement, typically used for electric guitar, bass, and handheld microphones. This lets performers move about the stage during the show or even go out into the audience without the worry of tripping over or disconnecting the cable.

Mixing Consoles

Mixing consoles are the heart of a sound reinforcement system. This is where the operator can adjust the volume and tone of each input, whether it is a vocalist's microphone or the signal from an electric bass, and mix, equalize and add effects to these sound sources. Multiple consoles can be used for different applications in a single sound reinforcement system. The Front of House (FOH) mixing console must be located where the operator can see the action on stage and hear the output of the loudspeaker system. Some venues with permanently installed systems such as religious facilities and theaters place the mixing console within an enclosed booth, but this approach is more common for broadcast and recording applications. This is far less common in live sound reproduction, since the engineer performs best when she can hear wthat the audience hears.

Large music productions often use a separate stage monitor mixing console which is dedicated to creating mixes for the performers' on-stage or in-ear monitors. These consoles are typically placed at the side of the stage so that the operator can communicate with the performers on stage. In cases where performers have to play at a venue that does not have a monitor engineer near the stage, the monitor mixing is done by the FOH engineer from the FOH console, which is located amongst the audience or at the back of the hall. This arrangement can be problematic because the performers end up having to request changes to the monitor mixes with "...hand signals and clever cryptic phrases" which may be misunderstood. The engineer also cannot hear the changes that she is applying to the monitors on stage, often resulting in a reduction of the quality of the onstage monitor mix.

Signal Processors

In the 2010s, small PA systems for venues such as bars and clubs are now available with features that were formerly only available on professional-level equipment, such as digital reverb effects, graphic equalizers, and, in some models, feedback prevention circuits which electronically sense and prevent feedback "howls" before they become a problem. Digital effects units may offer multiple pre-set and variable reverb, echo and related effects. Digital loudspeaker management systems offer sound engineers digital delay, limiting, crossover functions, EQ filters, compression and other functions in a single rack-mountable unit. In previous decades, sound engineers typically had to transport a substantial number of rack-mounted analog effects unit devices to accomplish these tasks.

Equalizers

Equalizers are electronic devices that allow audio engineers to control the tone of the sound in a channel or an entire mix. The bass and treble controls on a home stereo are a simple type of equalizer. Equalizers exist in pro sound reinforcement systems in two forms: graphic equalizers and parametric equalizers. Graphic equalizers have faders (vertical slide controls) which together resemble a frequency response curve plotted on a graph. The faders can be used to boost or cut specific frequency bands. Frequencies which are too weak can be boosted. Frequencies which are too loud, such as a "boomy" sounding bass drum, can be cut. Sound reinforcement systems typically use graphic equalizers with one-third octave frequency centers. These are typically used to equalize output signals going to the main loudspeaker system or the monitors on stage. Parametric equalizers are often built into each channel in mixing consoles and are also available as separate units. Parametric equalizers typically use knobs. The audio engineer can select which frequency band to cut or boost, and then use additional knobs to cut or boost this frequency range. Parametric equalizers first became popular in the 1970s and have remained the program equalizer of choice for many engineers since then.

Graphic equalizer

A high-pass (low-cut) and/or low-pass (high-cut) filter may also be included on equalizers or audio consoles. High-pass (low-cut) and low-pass (high-cut) filters restrict a given channel's bandwidth extremes. Cutting very low frequency sound signals (termed infrasonic, or *subsonic*, a misnomer) reduces the waste of amplifier power which does not produce audible sound and which moreover can be hard on the low-range speakers. A low-pass filter to cut ultrasonic energy is useful to prevent interference from radio frequencies, lighting control hum, or digital circuitry creeping into the power amplifiers. Such filters are often included with graphic and parametric equalizers to give the audio engineer full control of the frequency range. If their response is steep enough, high-pass filters and low-pass filters function as end-cut filters. A feedback suppressor is an automatically-adjusted band-reject or notch filter which includes a microprocessor to detect the onset of feedback "howls" and direct the filter to suppress the feedback by lowering the gain right at the offending frequency.

Compressors

Compressors are designed to help the audio engineer to manage the dynamic range of an audio signal, or a group of audio signals. A compressor accomplishes this by reducing the gain of a signal that is above a defined level (the threshold) by a defined amount (the ratio). Without this gain reduction, a signal that gets, say 10% louder as an input, will be 10% louder at the output. With the gain reduced, a signal that gets 10% louder at the input will be perhaps 3% louder at the output. Most compressors available are designed to allow the operator to select a ratio within a range typically between 1:1 and 20:1, with some allowing settings of up to ∞:1. A compressor with an infinite ratio is typically referred to as a *limiter*. The speed that the compressor adjusts the gain of the signal (called the attack) is typically adjustable as is the final output of the device.

A rack of electronic audio compressors

Compressor applications vary widely from objective system design criterion to subjective applications determined by variances in program material and engineer preference. Some system design criteria specify limiters for component protection and gain structure control. Artistic signal manipulation is a subjective technique widely utilized by mix engineers to improve clarity or to creatively alter the signal in relation to the program material. An example of artistic compression is the typical heavy compression used on the various components of a modern rock drum kit. The drums are processed to be perceived as sounding more punchy and full.

Effect processing rack-mounted units at the FOH position at an outdoor concert.

Noise Gates

A noise gate sets a threshold where if it is quieter it will not let the signal pass and if it is louder it opens the gate. A noise gate's function is in a sense the opposite to that of a compressor. Noise gates are useful for microphones which will pick up noise which is not relevant to the program, such as the hum of a miked electric guitar amplifier or the rustling of papers on a minister's lectern. Noise gates are also used to process the microphones placed near the drums of a drum kit in many hard rock and metal bands. Without a noise gate, the microphone for a specific instrument such as the floor tom will also pick up signal from nearby drums or cymbals. With a noise gate, the threshold of sensitivity for each microphone on the drum kit can be set so that only the direct strike and subsequent decay of the drum will be heard, not the nearby sounds.

Effects

Reverberation and delay effects are widely used in sound reinforcement systems to enhance the sound of the mix and create a desired artistic effect. Reverb and delay add a sense of spaciousness to the sound, imitating the sound of a singing voice or instrument in a large, reverberant hall. Modulation effects such as flanger, phaser, and chorus are also applied to some instruments. An exciter "livens up" the sound of audio signals by applying dynamic equalization, phase manipulation and harmonic synthesis of typically high frequency signals.

The appropriate type, variation, and level of effects is quite subjective and is often collectively determined by a production's engineer, artists, bandleader, music producer, or musical director. Reverb, for example, can give the effect of signal being present in anything from a small room to a massive hall, or even in a space that doesn't exist in the physical world. The use of reverb often goes unnoticed by the audience, as it often sounds more natural than if the signal was left "dry" (without effects). The use of effects in the reproduction of 2010-era pop music is often in an attempt to mimic the sound of the studio version of the artist's music in a live concert setting.

Feedback Suppressor

A feedback suppressor detects audio feedback and suppresses it typically by automatically inserting a notch filter into the signal path of the system, which prevents feedback "howls" from occurring. Audio feedback can create unwanted loud, noises which are disruptive to the performance, and which can damage performers' and audience members' ears and speakers. Audio feedback from microphones occurs when a microphone "hears" the sound it is picking up through the monitor speakers or the main speakers. While microphone audio feedback is almost universally regarded as a negative phenomenon, in hard rock and heavy metal music, electric guitarists purposely create guitar feedback to create unique, sustained sounds. This type of feedback is sought out by guitarists, so the sound engineer does not try to prevent it.

Power Amplifiers

A power amplifier boosts a low-voltage level signal and provides enough electrical power to drive a loudspeaker. All speakers, including headphones, require power amplification. Most professional audio amplifiers also provide protection from overdriven signals, short circuits across the output, and excessive temperature. A limiter is often used to protect loudspeakers and amplifiers from overload.

Like most sound reinforcement equipment products, professional amplifiers are designed to be mounted within standard 19-inch racks. Active loudspeakers feature internally mounted amplifiers that have been selected by the manufacturer to be a good amplifier for use with the given loudspeaker.

Three audio power amplifiers

Since amplifiers can generate a significant amount of heat, thermal dissipation is an important factor for operators to consider when mounting amplifiers into equipment racks. Many power amplifiers feature internal fans to draw air across their heat sinks. The heat sinks can become clogged with dust, which can adversely affect the cooling capabilities of the amplifier.

In the 1970s and 1980s, most PAs employed heavy Class AB amplifiers. In the late 1990s, power amplifiers in PA applications became lighter, smaller, more powerful, and more efficient, with the increasing use of switching power supplies and Class D amplifiers, which offered significant weight- and space-savings as well as increased efficiency. Often installed in railroad stations, stadia, and airports, Class D amplifiers can run with minimal additional cooling and with higher rack densities, compared to older amplifiers.

Digital loudspeaker management systems (DLMS) that combine digital crossover functions, compression, limiting, and other features in a single unit have become popular since their introduction. They are used to process the mix from the mixing console and route it to the various amplifiers. Systems may include several loudspeakers, each with its own output optimized for a specific range of frequencies (i.e. bass, midrange, and treble). Bi-amplifiication, tri-amplification, or quad-amplification of a sound reinforcement system with the aid of a DLMS results in a more efficient use of amplifier power by sending each amplifier only the frequencies appropriate for its respective loudspeaker. Most DLMS units that are designed for use by non-professionals have calibration and testing functions such as a pink noise generator coupled with a real-time analyzer to allow automated room equalization.

Output Transducers

Main Loudspeakers

A simple and inexpensive PA loudspeaker may have a single full-range loudspeaker driver, housed in a suitable enclosure. More elaborate, professional-caliber sound reinforcement loudspeakers may incorporate separate drivers to produce low, middle, and high frequency sounds. A crossover network routes the different frequencies to the appropriate drivers. In the 1960s, horn loaded theater loudspeakers and PA speakers were almost always "columns" of multiple drivers mounted in a vertical line within a tall enclosure. The 1970s to early 1980s was a period of innovation in loudspeaker design with many sound reinforcement companies designing their own speakers. The basic designs were based on commonly known designs and the speaker components were commercial speakers.

A large line array with separate subs and a smaller side fill line array.

The areas of innovation were in cabinet design, durability, ease of packing and transport, and ease of setup. This period also saw the introduction of the hanging or "flying" of main loudspeakers at large concerts. During the 1980s the large speaker manufacturers started producing standard products using the innovations of the 1970s. These were mostly smaller two way systems with 12", 15" or double 15" woofers and a high frequency driver attached to a high frequency horn. The 1980s also saw the start of loudspeaker companies focused on the sound reinforcement market. The 1990s saw the introduction of Line arrays, where long vertical arrays of loudspeakers with a smaller cabinet are used to increase efficiency and provide even dispersion and frequency response. This period also saw the introduction of inexpensive molded plastic speaker enclosures mounted on tripod stands. Many feature built-in power amplifiers which made them practical for non-professionals to set up and operate successfully. The sound quality available from these simple 'powered speakers' varies widely depending on the implementation.

Many sound reinforcement loudspeaker systems incorporate protection circuitry, preventing damage from excessive power or operator error. Positive temperature coefficient resistors, specialized current-limiting light bulbs, and circuit-breakers were used alone or in combination to reduce driver failures. During the same period, the professional sound reinforcement industry made the Neutrik Speakon NL4 and NL8 connectors the standard input connectors, replacing 1/4" jacks, XLR connectors, and Cannon multipin connectors which are all limited to a maximum of 15 amps of current. XLR connectors are still the standard input connector on active loudspeaker cabinets.

The three different types of transducers are subwoofers, compression drivers, and tweeters. They all feature the combination of a voicecoil, magnet, cone or diaphragm, and a frame or structure. Loudspeakers have a power rating (in watts) which indicates their maximum power capacity, to help users avoid overpowering them. Thanks to the efforts of the Audio Engineering Society (AES) and the loudspeaker industry group ALMA, power-handling specifications became more trustworthy, although adoption of the EIA-426-B standard is far from universal. Around the mid 1990s trapezoidal-shaped enclosures became popular as this shape allowed many of them to be easily arrayed together.

A number of companies are now making lightweight, portable speaker systems for small venues that route the low-frequency parts of the music (electric bass, bass drum, etc.) to a powered sub-woofer. Routing the low-frequency energy to a separate amplifier and subwoofer can substantially improve the bass-response of the system. Also, clarity may be enhanced, because low-frequency sounds take a great deal of power to amplify; with only a single amplifier for the entire sound spec-trum, the power-hungry low-frequency sounds can take a disproportionate amount of the sound system's power.

Professional sound reinforcement speaker systems often include dedicated hardware for safely "flying" them above the stage area, to provide more even sound coverage and to maximize sight lines within performance venues.

Monitor Loudspeakers

Monitor loudspeakers, also called "foldback" loudspeakers, are speaker cabinets which are used onstage to help performers to hear their singing or playing. As such, monitor speakers are pointed towards a performer or a section of the stage. They are generally sent a different mix of vocals or instruments than the mix that is sent to the main loudspeaker system. Monitor loudspeaker cabinets are often a wedge shape, directing their output upwards towards the performer when set on the floor of the stage. Two-way, dual driver designs with a speaker cone and a horn are common, as monitor loudspeakers need to be smaller to save space on the stage. These loudspeakers typically require less power and volume than the main loudspeaker system, as they only need to provide sound for a few people who are in relatively close proximity to the loudspeaker. Some manufacturers have designed loudspeakers for use either as a component of a small PA system or as a monitor loudspeaker. In the 2000s, a number of manufac-turers produced powered monitor speakers, which contain an integrated amplifier.

A JBL floor monitor speaker cabinet with a 12" (30 cm) woofer and a "bullet" tweeter. Most monitor cabinets have a metal grille or woven plastic mesh to protect the loudspeaker.

Using monitor speakers instead of in ear monitors typically results in an increase of stage volume, which can lead to more feedback issues and progressive hearing damage for the performers in front of them. The clarity of the mix for the performer on stage is also typically not as clear as they hear more extraneous noise from around them. The use of monitor loudspeakers, active (with an integrated amplifier) or passive, requires more cabling and gear on stage, resulting in an even more cluttered stage. These factors, amongst others, have led to the increasing popularity of in-ear monitors.

In-ear Monitors

In-ear monitors are headphones that have been designed for use as monitors by a live performer. They are either of a "universal fit" or "custom fit" design. The universal fit in ear monitors feature rubber or foam tips that can be inserted into virtually anybody's ear. Custom fit in ear monitors are created from an impression of the users ear that has been made by an audiologist. In-ear monitors are almost always used in conjunction with a wireless transmitting system, allowing the performer to freely move about the stage while maintaining their monitor mix.

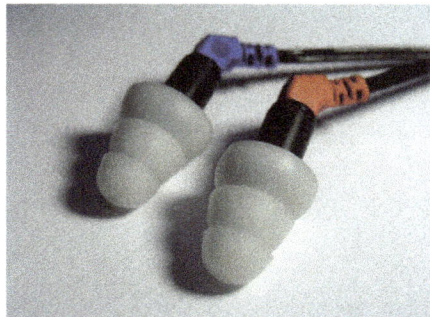

A pair of universal fit in-ear monitors. This particular model is the Etymotic ER-4S

In-ear monitors offer considerable isolation for the performer using them, meaning that the monitor engineer can craft a much more accurate and clear mix for the performer. With in-ear monitors, each performer can be sent their own customized mix; although this was also the case with monitor speakers, the in-ear monitors of one performer cannot be heard by the other musicians. A downside of this isolation is that the performer cannot hear the crowd or the comments other performers on stage that do not have microphones (e.g., if the bass player wishes to communicate to the drummer). This has been remedied by larger productions by setting up a pair of microphones on each side of the stage facing the audience that are mixed into the in-ear monitor sends.

Since their introduction in the mid-1980s, in-ear monitors have grown to be the most popular monitoring choice for large touring acts. The reduction or elimination of loudspeakers other than instrument amplifiers on stage has allowed for cleaner and less problematic mixing situations for both the front of house and monitor engineers. Audio feedback is greatly reduced and there is less sound reflecting off the back wall of the stage out into the audience, which affects the clarity of the mix the front of house engineer is attempting to create.

Applications

Sound reinforcement systems are used in a broad range of different settings, each of which poses different challenges.

Rental Systems

Audio visual (AV) rental systems have to be able to withstand heavy use, and even abuse from renters. For this reason, rental companies tend to own speaker cabinets which are heavily braced and protected with steel corners, and electronic equipment such as power amplifiers or effects are often mounted into protective road cases. As well, rental companies tend to select gear which has electronic protection features, such as speaker-protection circuitry and amplifier limiters.

As well, rental systems for non-professionals need to be easy to use and set up, and they must be easy to repair and maintain for the renting company. From this perspective, speaker cabinets need to have easy-to-access horns, speakers, and crossover circuitry, so that repairs or replacements can be made. Some rental companies often rent powered amplifier-mixers, mixers with onboard effects, and powered subwoofers for use by non-professionals, which are easier to set up and use.

Many touring acts and large venue corporate events will rent large sound reinforcement systems that typically include one or more audio engineers on staff with the renting company. In the case of rental systems for tours, there are typically several Engineers and Technicians from the Rental company that tour with the act to set up and calibrate the equipment for use by the band's production crew. The individual that actually mixes the act is often selected and provided by the band, as they are someone who has become familiar with the various aspects of the show and have worked with the act to establish a general idea of how they want the show to sound. The mixing engineer for an act sometimes also happens to be on staff with the rental company selected to provide the gear for the tour.

Live Music Clubs

Setting up sound reinforcement for live music clubs often poses unique challenges, because there is such a large variety of venues which are used as clubs, ranging from former warehouses or music theaters to small restaurants or basement pubs with concrete walls. In some cases, clubs are housed in multi-story venues with balconies or in "L"-shaped rooms, which makes it hard to get a consistent sound for all audience members. The solution is to use fill-in speakers to obtain good coverage, using a delay to ensure that the audience does not hear the same sound at different times.

A front-of-house sound engineer with a Digidesign D-Show Profile live digital mixer and a computer monitor.

Another problem with designing sound systems for live music clubs is that the sound system may need to be used for both prerecorded music played by DJs and live music. If the sound system is optimized for prerecorded DJ music, then it will not provide the appropriate sound qualities (or mixing and monitoring equipment) needed for live music, and vice versa. Lastly, live music clubs can be a hostile environment for sound gear, in that the air may be hot, humid, and smoky; in some clubs, keeping racks of power amplifiers cool may be a challenge. Often an air conditioned room just for the amplifiers is utilised.

Church Sound

Designing systems in churches and similar religious facilities often poses a challenge, because the speakers may have to be unobtrusive to blend in with antique woodwork and stonework. In some cases, audio designers have designed custom-painted speaker cabinets so that the speakers will blend in with the church architecture. Some church facilities, such as sanctuaries or chapels are long rooms with low ceilings, which means that additional fill-in speakers are needed throughout the room to give good coverage. An additional challenge with church SR systems is that, once installed, they are often operated by amateur volunteers from the congregation, which means that they must be easy to operate and troubleshoot.

Some mixing consoles designed for houses of worship have automatic mixers, which turn down unused channels to reduce noise, and automatic feedback elimination circuits which detect and notch out frequencies that are feeding back. These features may also be available in multi-function consoles used in convention facilities and multi-purpose venues.

Touring Systems

Touring sound systems have to be powerful and versatile enough to cover many different rooms, often being of many different sizes and shapes. They also need to use "field-replaceable" components such as speakers, horns, and fuses, which are easily accessible for repairs during a tour. Tour sound systems are often designed with substantial redundancy features, so that in the event of equipment failure or amplifier overheating, the system will continue to function. Touring systems for acts performing for crowds of a few thousand people and up are typically set up and operated by a team of technicians and engineers that travel with the talent to every show.

It is not uncommon for mainstream acts that are going to perform in mid to large venues during their tour to schedule one to two weeks of tech rehearsal with the entire concert system and production staff at hand. This allows the audio and lighting engineers to become familiar with the show and establish presets on their digital equipment for each part of the show, if needed. Many modern musical groups work with their Front of House and Monitor Mixing Engineers during this time to establish what their general idea is of how the show should sound, both for themselves on stage and for the audience. This often involves programming different effects and signal processing for use on specific songs in an attempt to make the songs sound somewhat similar to the studio versions. To manage a show with a lot of these types of changes, the mixing engineers for the show often choose to use a digital mixing console so that they can recall these many settings in between each song. This time is also used by the system technicians to get familiar with the specific combination of gear that is going to be used on the tour and how it acoustically responds during the show. These technicians remain busy during the show, making sure the SR system is operating properly and that the system is tuned correctly, as the acoustic response of a room will respond differently throughout the day depending on the temperature, humidity, and number of people in the room.

Weekend band PA systems are a niche market for touring SR gear. Weekend bands need systems that are small enough to fit into a minivan or a car trunk, and yet powerful enough to give adequate and even sound dispersion and vocal intelligibility in a noisy club or bar. As well, the systems need to be easy and quick to set up. Sound reinforcement companies have responded to this demand by offering equipment that fulfills multiple roles, such as powered mixers (a mixer with an integrated power amplifier and effects) and powered subwoofers (a subwoofer with an integrated power amplifier and crossover). These products minimize the amount of wiring connections that bands have to make to set up the system. Some subwoofers have speaker mounts built into the top, so that they can double as a base for the stand-mounted full-range PA speaker cabinets.

Live Theater

Sound for live theater, operatic theater, and other dramatic applications may pose problems similar to those of churches, in cases where a theater is an old heritage building where speakers and wiring may have to blend in with woodwork. The need for clear sight lines in some theaters may make the use of regular speaker cabinets unacceptable; instead, slim, low-profile speakers are often used instead.

In live theater and drama, performers move around onstage, which means that wireless microphones may have to be used. Wireless microphones need to be set up and maintained properly, to avoid interference and reception problems.

Some of the higher budget theater shows and musicals are mixed in surround sound live, often with the show's sound operator triggering sound effects that are being mixed with music and dialogue by the show's mixing engineer. These systems are usually much more extensive to design, typically involving a separate sets of speakers for different zones in the theater.

Classical Music and Opera

A subtle type of sound reinforcement called acoustic enhancement is used in some concert halls where classical music such as symphonies and opera is performed. Acoustic enhancement systems help give a more even sound in the hall and prevent "dead spots" in the audience seating area by "...augment[ing] a hall's intrinsic acoustic characteristics." The systems use "...an array of microphones connected to a computer [which is] connected to an array of loudspeakers." However, as concertgoers have become aware of the use of these systems, debates have arisen, because "...purists maintain that the natural acoustic sound of [Classical] voices [or] instruments in a given hall should not be altered."

Kai Harada's article *Opera's Dirty Little Secret* states that opera houses have begun using electronic acoustic enhancement systems "...to compensate for flaws in a venue's acoustical architecture." Despite the uproar that has arisen amongst operagoers, Harada points out that none of the opera houses using acoustic enhancement systems "...use traditional, Broadway-style sound reinforcement, in which most if not all singers are equipped with radio microphones mixed to a series of unsightly loudspeakers scattered throughout the theatre." Instead, most opera houses use the sound reinforcement system for acoustic enhancement, and for subtle boosting of offstage voices, onstage dialogue, and sound effects (e.g., church bells in Tosca or thunder in Wagnerian operas).

Acoustic enhancement systems include LARES (Lexicon Acoustic Reinforcement and Enhancement System) and SIAP, the System for Improved Acoustic Performance. These systems use microphones, computer processing "with delay, phase, and frequency-response changes", and then send the signal "... to a large number of loudspeakers placed in extremities of the performance venue." Another acoustic enhancement system, VRAS (Variable Room Acoustics System) uses "... different algorithms based on microphones placed around the room." The Deutsche Staatsoper in Berlin and the Hummingbird Centre in Toronto use a LARES system. The Ahmanson Theatre in Los Angeles, the Royal National Theatre in London, and the Vivian Beaumont Theater in New York City use the SIAP system.

Lecture Halls and Conference Rooms

Lecture halls and conference rooms pose the challenge of reproducing speech clearly to a large hall, which may have reflective, echo-producing surfaces. One issue with reproducing speech is that the microphone used to pick up the sound of an individual's voice may also pick up unwanted sounds, such as the rustling of papers on a podium. A more tightly directional microphone may help to reduce unwanted background noises. Another challenge with doing live sound for individuals who are speaking at a conference is that, in comparison with professional singers, individuals who are invited to speak at a forum may not be familiar with how microphones work. Some individuals may accidentally point the microphone towards a speaker or monitor speaker, which may cause audio feedback "howls". In some cases, when an individual who is speaking does not speak enough directly into the microphone, the audio engineer may ask the individual to wear a lavaliere microphone, which can be clipped onto a lapel. In some conferences, sound engineers have to provide microphones for a large number of people who are speaking, in the case of a panel conference or debate. In some cases, automatic mixers are used to control the levels of the microphones, and turn off the channels for microphones which are not being spoken into, to reduce unwanted background noise and reduce the likelihood of feedback.

Sports Sound Systems

Systems for outdoor sports facilities and ice rinks often have to deal with substantial echo, which can make speech unintelligible. Sports and recreational sound systems often face environmental challenges as well, such as the need for weather-proof outdoor speakers in outdoor stadiums and humidity- and splash-resistant speakers in swimming pools.

Setting Up and Testing

Large-scale sound reinforcement systems are designed, installed, and operated by audio engineers and audio technicians. During the design phase of a newly constructed venue, audio engineers work with architects and contractors, to ensure that the proposed design will accommodate the speakers and provide an appropriate space for sound technicians and the racks of audio equipment. Sound engineers will also provide advice on which audio components would best suit the space and its intended use, and on the correct placement and installation of these components. During the installation phase, sound engineers ensure that high-power electrical components are safely installed and connected and that ceiling or wall mounted speakers are properly mounted (or "flown") onto rigging. When the sound reinforcement components are installed, the sound engineers test and calibrate the system so that its sound production will be even across the frequency spectrum.

System Testing

A sound reinforcement system should be able to accurately reproduce a signal from its input, through any processing, to its output without any coloration or distortion. However, due to inconsistencies in venue sizes, shapes, building materials, and even crowd densities, this is not always possible without prior calibration of the system. This can be done in one of several ways. The oldest method of system calibration involves a set of healthy ears, test program material (i.e. music or speech), a graphic equalizer, and last but certainly not least, a familiarity with the proper (or desired) frequency response. One must then listen to the program material through the system, take note of any noticeable frequency changes or resonances, and subtly correct them using the equalizer. Experienced engineers typically use a specific playlist of music every time they calibrate a system that they have become very familiar with. This process is still done by many engineers, even when analysis equipment is used, as a final check of how the system sounds with music or speech playing through the system.

Another method of manual calibration requires a pair of high-quality headphones patched into the input signal *before* any processing (such as the pre-fade-listen of the test program input channel of the mixing console, or the headphone output of the CD player or tape deck). One can then use this direct signal as a near-perfect reference with which to find any differences in frequency response. This method may not be perfect, but it can be very helpful with limited resources or time, such as using pre-show music to correct for the changes in response caused by the arrival of a crowd. Because this is still a very subjective method of calibration, and because the human ear is so dynamic in its own response, the program material used for testing should be as similar as possible to that for which the system is being used.

Since the development of digital signal processing (DSP), there have been many pieces of equipment and computer software designed to shift the bulk of the work of system calibration from human auditory interpretation to software algorithms that run on microprocessors. One tool for calibrating a sound system using either DSP or Analog Signal Processing is a Real Time Analyzer (RTA). This tool is usually used by piping pink noise into the system and measuring the result with a special calibrated microphone connected to the RTA. Using this information, the system can be adjusted to help achieve the desired response. The displayed response from the RTA mic cannot be taken as a perfect representation of the room as the analysis will be different, sometimes drastically, when the mic is placed in different position in front of the system.

A Rane RA 27 hardware Real Time Analyzer underneath an Ashly Protea II 4.24C speaker processor (with RS-232 connection)

More recently, sound engineers have seen the introduction of dual "fft" (fast-fourier transform) based audio analysis software which allows an engineer to view not only frequency vs. amplitude (pitch vs. volume) information that an RTA provides, but also to see the same signals (sounds) in the time domain. This provides the engineer with much more meaningful data than an rta alone. Also, dual fft analysis allows one to compare the source signal with the output signal and view the difference. This is a very fast way to calibrate a system to sound as close as possible to the original source material. As with any such measurement tool, it must always be verified using actual human ears. Some DSP system processing devices have been designed for use by non-professionals that automatically make adjustments in the system EQ based upon what is being read from the RTA mic. These are practically never used by professionals, as they almost never calibrate the system as well as a professional audio engineer can manually.

Equipment Supply Stores

Professional audio stores sell microphones, speaker enclosures, monitor speakers, mixing boards and related equipment designed for use by audio engineers and technicians. Professional audio stores are also called "pro audio stores", "pro sound stores", "sound reinforcement" companies, "PA system companies" or "audio-visual companies", with the latter name being used when a store supplies a significant amount of video equipment for events, such as video projectors and screens. Stores often use the word "professional" in their name or the description of their store, to differentiate their stores from consumer electronics stores, which sell consumer-grade loudspeakers, home cinema equipment, and amplifiers, which are designed for private, in-home use.

References

- Ravaioli, Fawwaz T. Ulaby, Eric Michielssen, Umberto (2010). Fundamentals of applied electromagnetics (6th ed.). Boston: Prentice Hall. p. 13. ISBN 978-0-13-213931-1.

- International Union of Pure and Applied Chemistry (1993). Quantities, Units and Symbols in Physical Chemistry, 2nd edition, Oxford: Blackwell Science. ISBN 0-632-03583-8. pp. 14–15.

- Halliday, David; Robert Resnick; Kenneth S. Krane (1992). Physics. New York: John Wiley & Sons. ISBN 0-471-80457-6.

- Griffiths, David J. (1999). Introduction to Electrodynamics. Upper Saddle River, NJ: Prentice Hall. ISBN 0-13-805326-X.

- Hermann A. Haus; James R. Melcher (1989). Electromagnetic Fields and Energy. Englewood Cliffs, NJ: Prentice-Hall. ISBN 0-13-249020-X.

- Holler, F. James; Skoog, Douglas A.; Crouch, Stanley R. (2007). "Chapter 1". Principles of Instrumental Analysis (6th ed.). Cengage Learning. p. 9. ISBN 978-0-495-01201-6.

- S. Trolier-McKinstry (2008). "Chapter3: Crystal Chemistry of Piezoelectric Materials". In A. Safari; E.K. Akdoˇgan. Piezoelectric and Acoustic Materials for Transducer Applications. New York: Springer. ISBN 978-0-387-76538-9.

- Becker, Robert O.; Marino, Andrew A. (1982). "Chapter 4: Electrical Properties of Biological Tissue (Piezoelectricity)". Electromagnetism & Life. Albany, New York: State University of New York Press. ISBN 0-87395-560-9.

- Sinha, Dhiraj; Amaratunga, Gehan (2015). "Electromagnetic Radiation Under Explicit symmetry Breaking,". Physical Review Letters. 114 (14): 147701. Bibcode:2015PhRvL.114n7701S. doi:10.1103/physrevlett.114.147701. PMID 25910163.

- Li, Xiaofeng; Strezov, Vladimir (2014). "Modelling piezoelectric energy harvesting potential in an educational building". Energy Conversion and Management. 85: 435–442. doi:10.1016/j.enconman.2014.05.096.

- Erhart, Jiří. "Piezoelectricity and ferroelectricity: Phenomena and properties" (PDF). Department of Physics, Technical University of Liberec. Archived from the original on May 8, 2014.

- Katzir, S. (2012). "Who knew piezoelectricity? Rutherford and Langevin on submarine detection and the invention of sonar". Notes Rec. R. Soc. 66 (2): 141–157. doi:10.1098/rsnr.2011.0049.

- Moubarak, P.; et al. (2012). "A Self-Calibrating Mathematical Model for the Direct Piezoelectric Effect of a New MEMS Tilt Sensor". IEEE Sensors Journal. 12 (5): 1033–1042. doi:10.1109/jsen.2011.2173188

- Sound Equipment - Loudspeakers, Amplifiers, Signal Processors, Mixers, Music Source & Microphones. Retrieved on 2011-12-11.

7

Wave Propagation: Pressure Levels

The manner in which waves travel is studied as wave propagation. Some of the topics listed in this chapter are sound pressure, absolute threshold of hearing and interference. The chapter on wave propagation offers an insightful focus, keeping in mind the complex subject matter.

Sound Pressure

Sound pressure or acoustic pressure is the local pressure deviation from the ambient (average, or equilibrium) atmospheric pressure, caused by a sound wave. In air, sound pressure can be measured using a microphone, and in water with a hydrophone. The SI unit of sound pressure is the pascal (Pa).

Mathematical Definition

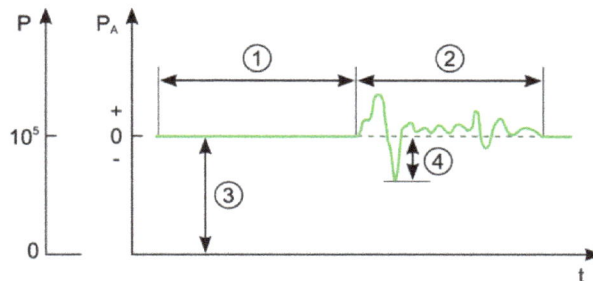

Sound pressure diagram:

1. silence;

2. audible sound;

3. atmospheric pressure;

4. sound pressure

A sound wave in a transmission medium causes a deviation (sound pressure, a *dynamic* pressure) in the local ambient pressure, a *static* pressure. Sound pressure, denoted p, is defined by

$$p_{total} = p_{stat} + p,$$

where

- p_{total} is the total pressure;

- p_{stat} is the static pressure.

Sound Measurements

Sound Intensity

In a sound wave, the complementary variable to sound pressure is the particle velocity. Together they determine the sound intensity of the wave. *Sound intensity*, denoted I and measured in W·m⁻² in SI units, is defined by

$$\mathbf{I} = p\mathbf{v},$$

where

- p is the sound pressure;
- v is the particle velocity.

Acoustic Impedance

Acoustic impedance, denoted Z and measured in Pa·m⁻³·s in SI units, is defined by

$$Z(s) = \frac{\hat{p}(s)}{\hat{Q}(s)},$$

where

- $\hat{p}(s)$ is the Laplace transform of sound pressure;
- $\hat{Q}(s)$ is the Laplace transform of sound volume flow rate.

Specific acoustic impedance, denoted z and measured in Pa·m⁻¹·s in SI units, is defined by

$$z(s) = \frac{\hat{p}(s)}{\hat{v}(s)},$$

where

- $\hat{p}(s)$ is the Laplace transform of sound pressure;
- $\hat{v}(s)$ is the Laplace transform of particle velocity.

Particle Displacement

The *particle displacement* of a *progressive sine wave* is given by

$$\delta(\mathbf{r},t) = \delta_{\mathrm{m}} \cos(\mathbf{k} \cdot \mathbf{r} - \omega t + \varphi_{\delta,0}),$$

where

- δ_{m} is the amplitude of the particle displacement;
- $\varphi_{\delta,0}$ is the phase shift of the particle displacement;
- k is the angular wavevector;
- ω is the angular frequency.

It follows that the particle velocity and the sound pressure along the direction of propagation of the sound wave x are given by

$$v(\mathbf{r},t) = \frac{\partial \delta}{\partial t}(\mathbf{r},t) = \omega \delta_m \cos\left(\mathbf{k} \cdot \mathbf{r} - \omega t + \varphi_{\delta,0} + \frac{\pi}{2}\right) = v_m \cos(\mathbf{k} \cdot \mathbf{r} - \omega t + \varphi_{v,0}),$$

$$p(\mathbf{r},t) = -\rho c^2 \frac{\partial \delta}{\partial x}(\mathbf{r},t) = \rho c^2 k_x \delta_m \cos\left(\mathbf{k} \cdot \mathbf{r} - \omega t + \varphi_{\delta,0} + \frac{\pi}{2}\right) = p_m \cos(\mathbf{k} \cdot \mathbf{r} - \omega t + \varphi_{p,0}),$$

where

- v_m is the amplitude of the particle velocity;
- $\varphi_{v,0}$ is the phase shift of the particle velocity;
- p_m is the amplitude of the acoustic pressure;
- $\varphi_{p,0}$ is the phase shift of the acoustic pressure.

Taking the Laplace transforms of v and p with respect to time yields

$$\hat{v}(\mathbf{r},s) = v_m \frac{s \cos \varphi_{v,0} - \omega \sin \varphi_{v,0}}{s^2 + \omega^2},$$

$$\hat{p}(\mathbf{r},s) = p_m \frac{s \cos \varphi_{p,0} - \omega \sin \varphi_{p,0}}{s^2 + \omega^2}.$$

Since $\varphi_{v,0} = \varphi_{p,0}$, the amplitude of the specific acoustic impedance is given by

$$z_m(\mathbf{r},s) = |z(\mathbf{r},s)| = \left|\frac{\hat{p}(\mathbf{r},s)}{\hat{v}(\mathbf{r},s)}\right| = \frac{p_m}{v_m} = \frac{\rho c^2 k_x}{\omega}.$$

Consequently, the amplitude of the particle displacement is related to that of the acoustic velocity and the sound pressure by

$$\delta_m = \frac{v_m}{\omega},$$

$$\delta_m = \frac{p_m}{\omega z_m(\mathbf{r},s)}.$$

Inverse-proportional Law

When measuring the sound pressure created by an object, it is important to measure the distance from the object as well, since the sound pressure of a *spherical* sound wave decreases as $1/r$ from the centre of the sphere (and not as $1/r^2$, like the sound intensity):

$$p(r) \propto \frac{1}{r}.$$

This relationship is an *inverse-proportional law*.

If the sound pressure p_1 is measured at a distance r_1 from the centre of the sphere, the sound pressure p_2 at another position r_2 can be calculated:

$$p_2 = \frac{r_1}{r_2} p_1.$$

The inverse-proportional law for sound pressure comes from the inverse-square law for sound intensity:

$$I(r) \propto \frac{1}{r^2}.$$

Indeed,

$$I(r) = p(r)v(r) = p(r)[p * z^{-1}](r) \propto p^2(r),$$

where

- $*$ is the convolution operator;

- z^{-1} is the convolution inverse of the specific acoustic impedance,

hence the inverse-proportional law:

$$p(r) \propto \frac{1}{r}.$$

The sound pressure may vary in direction from the centre of the sphere as well, so measurements at different angles may be necessary, depending on the situation. An obvious example of a sound source whose spherical sound wave varies in level in different directions is a bullhorn.

Sound Pressure Level

Sound pressure level (SPL) or acoustic pressure level is a logarithmic measure of the effective pressure of a sound relative to a reference value. Sound pressure level, denoted L_p and measured in dB, is defined by

$$L_p = \ln\left(\frac{p}{p_0}\right)\text{Np} = 2\log_{10}\left(\frac{p}{p_0}\right)\text{B} = 20\log_{10}\left(\frac{p}{p_0}\right)\text{dB},$$

where

- p is the root mean square sound pressure;

- p_0 is the *reference sound pressure*;

- $1\,\text{Np} = 1$ is the neper;

- $1\,\text{B} = (1/2)\ln(10)$ is the bel;

- $1\,\text{dB} = (1/20)\ln(10)$ is the decibel.

The commonly used reference sound pressure in air is

$$p_0 = 20\,\mu\text{Pa},$$

which is often considered as the threshold of human hearing (roughly the sound of a mosquito flying 3 m away). The proper notations for sound pressure level using this reference are $L_{p/(20\,\mu\text{Pa})}$ or L_p (re 20 µPa), but the suffix notations dB SPL, dB(SPL), dBSPL, or dB_{SPL} are very common, even if they are not accepted by the SI.

Most sound level measurements will be made relative to this reference, meaning 1 Pa will equal an SPL of 94 dB. In other media, such as underwater, a reference level of 1 µPa is used. These references are defined in ANSI S1.1-1994.

Examples

The lower limit of audibility is defined as SPL of 0 dB, but the upper limit is not as clearly defined. While 1 atm (194 dB Peak or 191 dB SPL) is the largest pressure variation an undistorted sound wave can have in Earth's atmosphere, larger sound waves can be present in other atmospheres or other media such as under water, or through the Earth.

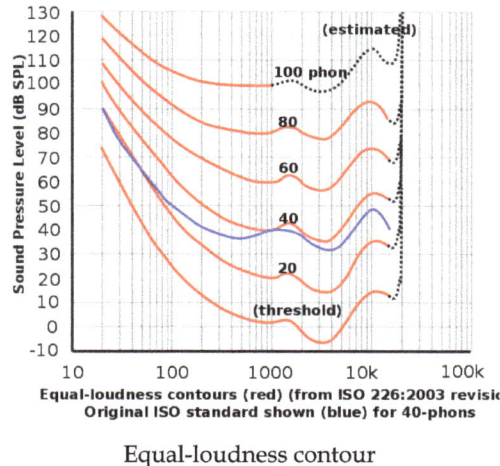

Equal-loudness contour

Ears detect changes in sound pressure. Human hearing does not have a flat spectral sensitivity (frequency response) relative to frequency versus amplitude. Humans do not perceive low- and high-frequency sounds as well as they perceive sounds between 3,000 and 4,000 Hz, as shown in the equal-loudness contour. Because the frequency response of human hearing changes with amplitude, three weightings have been established for measuring sound pressure: A, B and C. A-weighting applies to sound pressures levels up to 55 dB, B-weighting applies to sound pressures levels between 55 dB and 85 dB, and C-weighting is for measuring sound pressure levels above 85 dB.

In order to distinguish the different sound measures a suffix is used: A-weighted sound pressure level is written either as dB_A or L_A. B-weighted sound pressure level is written either as dB_B or L_B, and C-weighted sound pressure level is written either as dB_C or L_C. Unweighted sound pressure level is called "linear sound pressure level" and is often written as dB_L or just L. Some sound measuring instruments use the letter "Z" as an indication of linear SPL.

Distance

The distance of the measuring microphone from a sound source is often omitted when SPL measurements are quoted, making the data useless. In the case of ambient environmental measurements of "background" noise, distance need not be quoted as no single source is present, but when measuring the noise level of a specific piece of equipment the distance should always be stated. A distance of one metre (1 m) from the source is a frequently used standard distance. Because of the effects of reflected noise within a closed room, the use of an anechoic chamber allows for sound to be comparable to measurements made in a free field environment.

According to the inverse proportional law, when sound level L_{p1} is measured at a distance r_1, the sound level L_{p2} at the distance r_2 is

$$L_{p_2} = L_{p_1} + 20\log_{10}\left(\frac{r_1}{r_2}\right)\text{dB}.$$

Multiple Sources

The formula for the sum of the sound pressure levels of n incoherent radiating sources is

$$L_{\Sigma} = 10\log_{10}\left(\frac{p_1^2 + p_2^2 + \ldots + p_n^2}{p_0^2}\right)\text{dB} = 10\log_{10}\left[\left(\frac{p_1}{p_0}\right)^2 + \left(\frac{p_2}{p_0}\right)^2 + \ldots + \left(\frac{p_n}{p_0}\right)^2\right]\text{dB}.$$

Inserting the formulas

$$\left(\frac{p_i}{p_0}\right)^2 = 10^{\frac{L_i}{10\text{dB}}}, \quad i = 1,2,\ldots,n,$$

in the formula for the sum of the sound pressure levels yields

$$L_{\Sigma} = 10\log_{10}\left(10^{\frac{L_1}{10\text{dB}}} + 10^{\frac{L_2}{10\text{dB}}} + \ldots + 10^{\frac{L_n}{10\text{dB}}}\right)\text{dB}.$$

Absolute Threshold of Hearing

The absolute threshold of hearing (ATH) is the minimum sound level of a pure tone that an average human ear with normal hearing can hear with no other sound present. The absolute threshold relates to the sound that can just be heard by the organism. The absolute threshold is not a discrete point, and is therefore classed as the point at which a sound elicits a response a specified percentage of the time. This is also known as the auditory threshold.

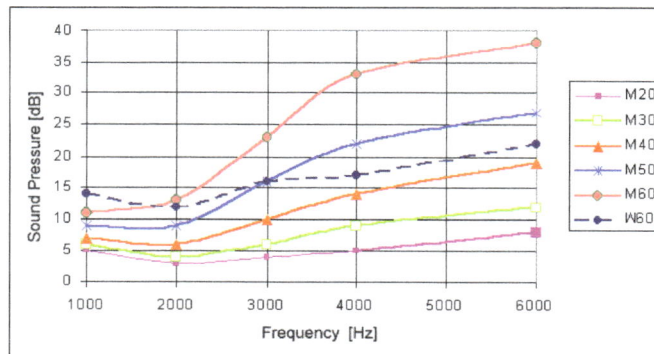

Thresholds of hearing for male (M) and female (W) subjects between the ages of 20 and 60

The threshold of hearing is generally reported as the RMS sound pressure of 20 micropascals, corresponding to a sound intensity of 0.98 pW/m² at 1 atmosphere and 25 °C. It is approximately the quietest sound a young human with undamaged hearing can detect at 1,000 Hz. The threshold of hearing is frequency-dependent and it has been shown that the ear's sensitivity is best at frequencies between 1 kHz and 5 kHz, where the threshold reaches as low as −9 dB SPL.

Psychophysical Methods for Measuring Thresholds

Measurement of the absolute hearing threshold provides some basic information about our auditory system. The tools used to collect such information are called psychophysical methods. Through these, the perception of a physical stimulus (sound) and our psychological response to the sound is measured.

Several psychophysical methods can measure absolute threshold. These vary, but certain aspects are identical. Firstly, the test defines the stimulus and specifies the manner in which the subject should respond. The test presents the sound to the listener and manipulates the stimulus level in a predetermined pattern. The absolute threshold is defined statistically, often as an average of all obtained hearing thresholds.

Some procedures use a series of trials, with each trial using the 'single-interval "yes"/"no" paradigm'. This means that sound may be present or absent in the single interval, and the listener has to say whether he thought the stimulus was there. When the interval does not contain a stimulus, it is called a "catch trial".

Classical Methods

Classical methods date back to the 19th century and were first described by Gustav Theodor Fechner in his work *Elements of Psychophysics*. Three methods are traditionally used for testing a subject's perception of a stimulus: the method of limits, the method of constant stimuli, and the method of adjustment.

Method of Limits

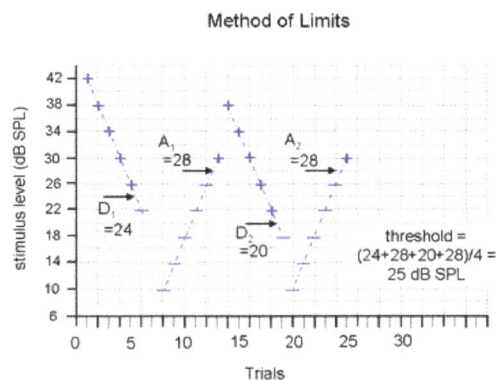

Series of descending and ascending runs in Method of Limits

In the method of limits, the tester controls the level of the stimuli. Single-interval *yes/no* paradigm' is used, but there are no catch trials.

The trial uses several series of descending and ascending runs.

The trial starts with the descending run, where a stimulus is presented at a level well above the expected threshold. When the subject responds correctly to the stimulus, the level of intensity of the sound is decreased by a specific amount and presented again. The same pattern is repeated until the subject stops responding to the stimuli, at which point the descending run is finished.

In the ascending run, which comes after, the stimulus is first presented well below the threshold and then gradually increased in two decibel (dB) steps until the subject responds.

As there are no clear margins to 'hearing' and 'not hearing', the threshold for each run is determined as the midpoint between the last audible and first inaudible level.

The subject's absolute hearing threshold is calculated as the mean of all obtained thresholds in both ascending and descending runs.

There are several issues related to the method of limits. First is anticipation, which is caused by the subject's awareness that the turn-points determine a change in response. Anticipation produces better ascending thresholds and worse descending thresholds.

Habituation creates completely opposite effect, and occurs when the subject becomes accustomed to responding either "yes" in the descending runs and/or "no" in the ascending runs. For this reason, thresholds are raised in ascending runs and improved in descending runs.

Another problem may be related to step size. Too large a step compromises accuracy of the measurement as the actual threshold may be just between two stimulus levels.

Finally, since the tone is always present, "yes" is always the correct answer.

Method of Constant Stimuli

In the method of constant stimuli, the tester sets the level of stimuli and presents them at completely random order.

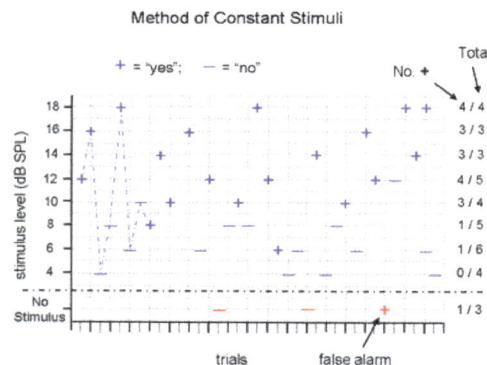

Subject responding "yes"/"no" after each presentation

Thus, there are no ascending or descending trials.

The subject responds "yes"/"no" after each presentation.

The stimuli are presented many times at each level and the threshold is defined as the stimulus level at which the subject scored 50% correct. "Catch" trials may be included in this method.

Method of constant stimuli has several advantages over the method of limits. Firstly, the random order of stimuli means that the correct answer cannot be predicted by the listener. Secondarily, as the tone may be absent (catch trial), "yes" is not always the correct answer. Finally, catch trials help to detect the amount of a listener's guessing.

The main disadvantage lies in the large number of trials needed to obtain the data, and therefore time required to complete the test.

Method of Adjustment

Method of adjustment shares some features with the method of limits, but differs in others. There are descending and ascending runs and the listener knows that the stimulus is always present.

Method of Adjustment

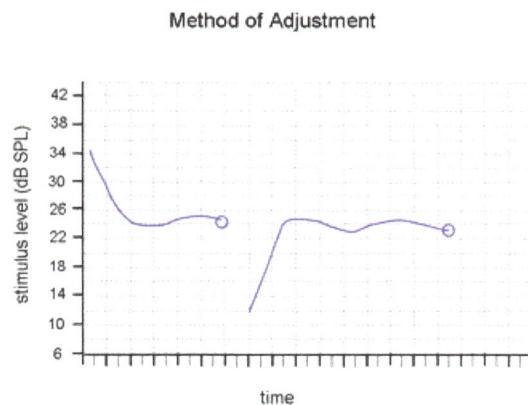

The subject reduces or increase the level of the tone

However, unlike in the method of limits, here the stimulus is controlled by the listener. The subject reduces the level of the tone until it cannot be detected anymore, or increases until it can be heard again.

The stimulus level is varied continuously via a dial and the stimulus level is measured by the tester at the end. The threshold is the mean of the just audible and just inaudible levels.

Also this method can produce several biases. To avoid giving cues about the actual stimulus level, the dial must be unlabeled. Apart from already mentioned anticipation and habituation, stimulus persistence (preservation) could influence the result from the method of adjustment.

In the descending runs, the subject may continue to reduce the level of the sound as if the sound was still audible, even though the stimulus is already well below the actual hearing threshold.

In contrast, in the ascending runs, the subject may have persistence of the absence of the stimulus until the hearing threshold is passed by certain amount.

Modified Classical Methods

Forced-choice Methods

Two intervals are presented to a listener, one with a tone and one without a tone. Listener must decide which interval had the tone in it. The number of the intervals can be increased, but this may cause problems to the listener who has to remember which interval contained the tone.

Adaptive Methods

Unlike the classical methods, where the pattern for changing the stimuli is preset, in adaptive methods the subject's response to the previous stimuli determines the level at which a subsequent stimulus is presented.

Staircase' Methods (Up-down Methods)

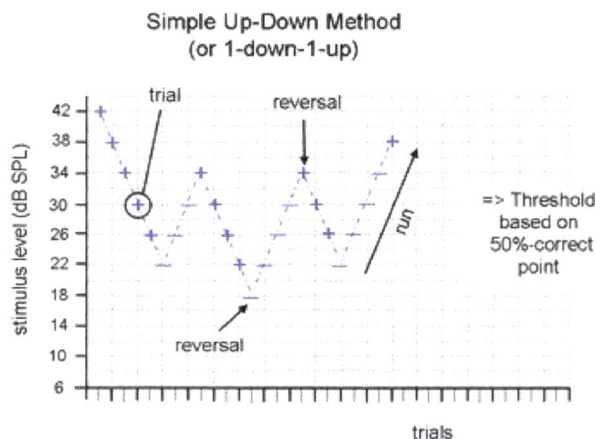

Series of descending and ascending trials runs and turning points

The simple '1-down-1-up' method consists of series of descending and ascending trials runs and turning points (reversals). The stimulus level is increased if the subject does not respond and decreased when a response occurs.

> Similarly, as in the method of limits, the stimuli are adjusted in predetermined steps. After obtaining from six to eight reversals, the first one is discarded and the threshold is defined as the average of the midpoints of the remaining runs. Experiments showed that this method provides only 50% accuracy.

> To produce more accurate results, this simple method can be further modified by increasing the size of steps in the descending runs, e.g. '2-down-1-up method', '3-down-1-up methods'.

Bekesy's Tracking Method

Bekesy's method contains some aspects of classical methods and staircase methods. The level of the stimulus is automatically varied at a fixed rate. The subject is asked to press a button when the stimulus is detectable.

Bekesy's Tracking Method

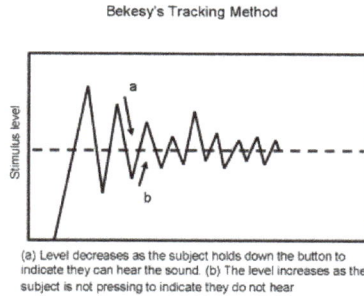

(a) Level decreases as the subject holds down the button to indicate they can hear the sound. (b) The level increases as the subject is not pressing to indicate they do not hear

The threshold being tracked by the listener

Once the button is pressed, the level is automatically decreased by the motor-driven attenuator and increased when the button is not pushed. The threshold is thus tracked by the listeners, and calculated as the mean of the midpoints of the runs as recorded by the automat.

Hysteresis Effect

Hysteresis can be defined roughly as 'the lagging of an effect behind its cause'. When measuring hearing thresholds it is always easier for the subject to follow a tone that is audible and decreasing in amplitude than to detect a tone that was previously inaudible.

This is because 'top-down' influences mean that the subject expects to hear the sound and is, therefore, more motivated with higher levels of concentration.

The 'bottom-up' theory explains that unwanted external (from the environment) and internal (e.g., heartbeat) noise results in the subject only responding to the sound if the signal to noise ratio is above a certain point.

In practice this means that when measuring threshold with sounds decreasing in amplitude, the point at which the sound becomes inaudible is always lower than the point at which it returns to audibility. This phenomenon is known as the 'hysteresis effect'.

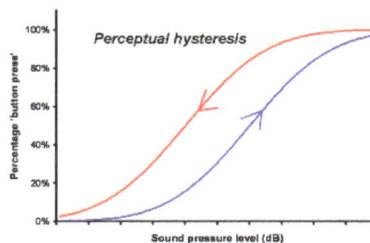

Descending runs give better hearing thresholds than ascending runs

Psychometric Function of Absolute Hearing Threshold

Psychometric function 'represents the probability of a certain listener's response as a function of the magnitude of the particular sound characteristic being studied'.

To give an example, this could be the probability curve of the subject detecting a sound being presented as a function of the sound level. When the stimulus is presented to the listener one would expect that the sound would either be audible or inaudible, resulting in a 'doorstep' function. In reality a grey area exists where the listener is uncertain as to whether they have actually heard the sound or not, so their responses are inconsistent, resulting in a psychometric function.

The psychometric function is a sigmoid function characterised by being 's' shaped in its graphical representation.

Minimal Audible Field (MAF) Vs Minimal Audible Pressure (MAP)

Two methods can be used to measure the minimal audible stimulus and therefore the absolute threshold of hearing. Minimal audible field involves the subject sitting in a sound field and stimulus being presented via a loudspeaker. The sound level is then measured at the position of the subjects head with the subject not in the sound field. Minimal audible pressure involves presenting stimuli via headphones or earphones and measuring sound pressure in the subject's ear canal using a very small probe microphone. The two different methods produce different thresholds and minimal audible field thresholds are often 6 to 10 dB better than minimal audible pressure thresholds. It is thought that this difference is due to:

- monaural vs binaural hearing. With minimal audible field both ears are able to detect the stimuli but with minimal audible pressure only one ear is able to detect the stimuli. Binaural hearing is more sensitive than monaural hearing.

- physiological noises heard when ear is occluded by an earphone during minimal audible pressure measurements. When the ear is covered the subject hears body noises, such as heart beat, and these may have a masking effect.

Minimal audible field and minimal audible pressure are important when considering calibration issues and they also illustrate that the human hearing is most sensitive in the 2–5 kHz range.

Temporal Summation

Temporal summation is the relationship between stimulus duration and intensity when the presentation time is less than 1 second. Auditory sensitivity changes when the duration of a sound becomes less than 1 second. The threshold intensity decreases by about 10 dB when the duration of a tone burst is increased from 20 to 200 ms.

For example, suppose that the quietest sound a subject can hear is 16 dB SPL if the sound is presented at a duration of 200 ms. If the same sound is then presented for a duration of only 20 ms, the quietest sound that can now be heard by the subject goes up to 26 dB SPL. In other words, if a signal is shortened by a factor of 10 then the level of that signal must be increased by as much as 10 dB to be heard by the subject.

The ear operates as an energy detector that samples the amount of energy present within a certain time frame. A certain amount of energy is needed within a time frame to reach the threshold. This can be done by using a higher intensity for less time or by using a lower intensity for more time. Sensitivity to sound improves as the signal duration increases up to about 200 to 300 ms, after that the threshold remains constant.

The timpani of the ear operates more as a sound pressure sensor. Also a microphone works the same way and is not sensitive to sound intensity.

Interference (Wave Propagation)

In physics, interference is a phenomenon in which two waves superpose to form a resultant wave of greater, lower, or the same amplitude. Interference usually refers to the interaction of waves that are correlated or coherent with each other, either because they come from the same source or because they have the same or nearly the same frequency. Interference effects can be observed with all types of waves, for example, light, radio, acoustic, surface water waves or matter waves.

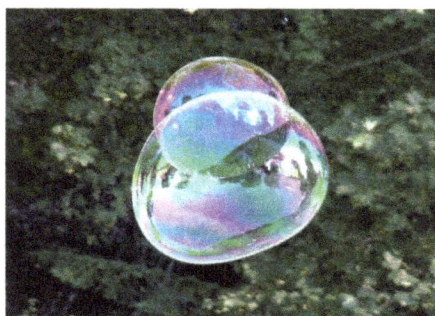

The iridescence of soap bubbles is due to thin-film interference.

Mechanism

The principle of superposition of waves states that when two or more propagating waves of same type are incident on the same point, the resultant amplitude at that point is equal to the vector sum of the amplitudes of the individual waves. If a crest of a wave meets a crest of another wave of the same frequency at the same point, then the amplitude is the sum of the individual amplitudes—this is constructive interference. If a crest of one wave meets a trough of another wave, then the amplitude is equal to the difference in the individual amplitudes—this is known as destructive interference.

Interference of left traveling (green) and right traveling (blue) waves in one dimension, resulting in final (red) wave

Constructive interference occurs when the phase difference between the waves is a multiple of 2π, whereas destructive interference occurs when the difference is an odd multiple of π. If the difference between the phases is intermediate between these two extremes, then the magnitude of the displacement of the summed waves lies between the minimum and maximum values.

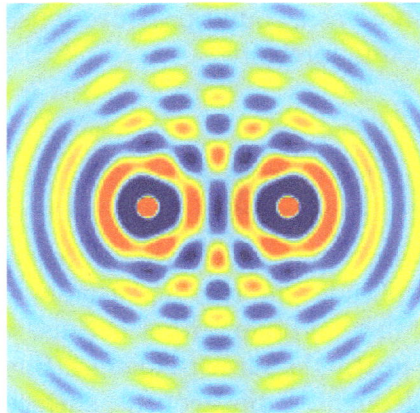

Interference of waves from two point sources.

Consider, for example, what happens when two identical stones are dropped into a still pool of water at different locations. Each stone generates a circular wave propagating outwards from the point where the stone was dropped. When the two waves overlap, the net displacement at a particular point is the sum of the displacements of the individual waves. At some points, these will be in phase, and will produce a maximum displacement. In other places, the waves will be in anti-phase, and there will be no net displacement at these points. Thus, parts of the surface will be stationary—these are seen in the figure above and to the right as stationary blue-green lines radiating from the center.

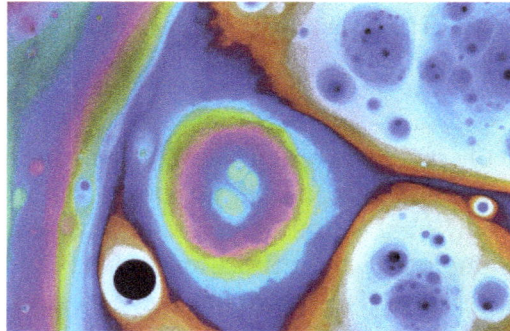

A magnified image of a coloured interference pattern in a soap film. The "black holes" are areas of almost total destructive interference, (antiphase).

Between Two Plane Waves

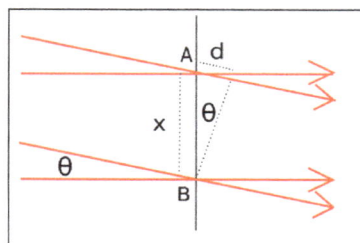

Geometrical arrangement for two plane wave interference

A simple form of interference pattern is obtained if two plane waves of the same frequency intersect at an angle. Interference is essentially an energy redistribution process. The energy which is lost at the destructive interference is regained at the constructive interference. One wave is travelling horizontally, and the other is travelling downwards at an angle θ to the first wave. Assuming that the two waves are in phase at the point B, then the relative phase changes along the x-axis. The phase difference at the point A is given by

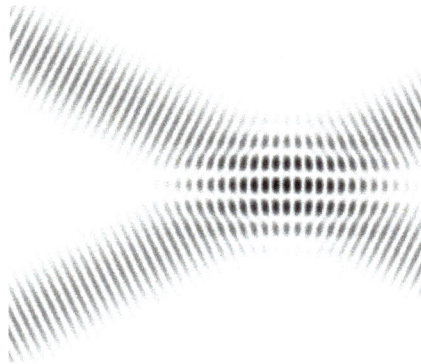

Interference fringes in overlapping plane waves

$$\ddot{A}\varphi = \frac{2\pi d}{\lambda} = \frac{2\pi x \sin\theta}{\lambda}$$

It can be seen that the two waves are in phase when

$$\frac{x\sin\theta}{\lambda} = 0, \pm 1, \pm 2, \dots,$$

and are half a cycle out of phase when

$$\frac{x\sin\theta}{\lambda} = \pm\frac{1}{2}, \pm\frac{3}{2}, \dots$$

Constructive interference occurs when the waves are in phase, and destructive interference when they are half a cycle out of phase. Thus, an interference fringe pattern is produced, where the separation of the maxima is

$$d_f = \frac{\lambda}{\sin\theta}$$

and d_f is known as the fringe spacing. The fringe spacing increases with increase in wavelength, and with decreasing angle θ.

The fringes are observed wherever the two waves overlap and the fringe spacing is uniform throughout.

Between Two Spherical Waves

A point source produces a spherical wave. If the light from two point sources overlaps, the interference pattern maps out the way in which the phase difference between the two waves varies in space. This depends on the wavelength and on the separation of the point sources. The figure to the right shows interference between two spherical waves. The wavelength increases from top to bottom, and the distance between the sources increases from left to right.

When the plane of observation is far enough away, the fringe pattern will be a series of almost straight lines, since the waves will then be almost planar.

Optical interference between two point sources for different wavelengths and source separations

Multiple Beams

Interference occurs when several waves are added together provided that the phase differences between them remain constant over the observation time.

It is sometimes desirable for several waves of the same frequency and amplitude to sum to zero (that is, interfere destructively, cancel). This is the principle behind, for example, 3-phase power and the diffraction grating. In both of these cases, the result is achieved by uniform spacing of the phases.

It is easy to see that a set of waves will cancel if they have the same amplitude and their phases are spaced equally in angle. Using phasors, each wave can be represented as $Ae^{i\varphi_n}$ for N waves from $n = 0$ to $n = N - 1$, , where

$$\varphi_n - \varphi_{n-1} = \frac{2\pi}{N}.$$

To show that

$$\sum_{n=0}^{N-1} Ae^{i\varphi_n} = 0$$

one merely assumes the converse, then multiplies both sides by $e^{i\frac{2\pi}{N}}$

The Fabry–Pérot interferometer uses interference between multiple reflections.

A diffraction grating can be considered to be a multiple-beam interferometer, since the peaks which it produces are generated by interference between the light transmitted by each of the elements in the grating.

Optical Interference

Creation of interference fringes by an optical flat on a reflective surface. Light rays from a monochromatic source pass through the glass and reflect off both the bottom surface of the flat and the supporting surface. The tiny gap between the surfaces means the two reflected rays have different path lengths and interfere when they combine. At locations (b) where the path difference is an even multiple of λ/2, the waves reinforce. At locations (a) where the path difference is an odd multiple of λ/2 the waves cancel. Since the gap between the surfaces varies slightly in width at different points, a series of alternating bright and dark bands are seen.

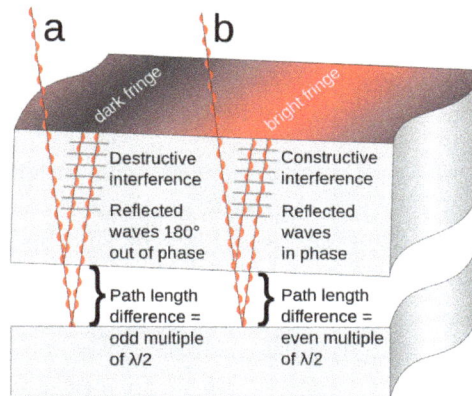

Because the frequency of light waves (~10^{14} Hz) is too high to be detected by currently available detectors, it is possible to observe only the intensity of an optical interference pattern. The intensity of the light at a given point is proportional to the square of the average amplitude of the wave. This can be expressed mathematically as follows. The displacement of the two waves at a point r is:

$$U_1(\mathbf{r},t) = A_1(\mathbf{r})e^{i[\varphi_1(\mathbf{r})-\omega t]}$$

$$U_2(\mathbf{r},t) = A_2(\mathbf{r})e^{i[\varphi_2(\mathbf{r})-\omega t]}$$

where A represents the magnitude of the displacement, φ represents the phase and ω represents the angular frequency.

The displacement of the summed waves is

$$U(\mathbf{r},t) = A_1(\mathbf{r})e^{i[\varphi_1(\mathbf{r})-\omega t]} + A_2(\mathbf{r})e^{i[\varphi_2(\mathbf{r})-\omega t]}$$

The intensity of the light at r is given by

$$I(\mathbf{r}) = \int U(\mathbf{r},t)U^*(\mathbf{r},t)dt \propto A_1^2(\mathbf{r}) + A_2^2(\mathbf{r}) + 2A_1(\mathbf{r})A_2(\mathbf{r})\cos[\varphi_1(\mathbf{r})-\varphi_2(\mathbf{r})]$$

This can be expressed in terms of the intensities of the individual waves as

$$I(\mathbf{r}) = I_1(\mathbf{r}) + I_2(\mathbf{r}) + 2\sqrt{I_1(\mathbf{r})I_2(\mathbf{r})}\,\cos[\varphi_1(\mathbf{r})-\varphi_2(\mathbf{r})]$$

Thus, the interference pattern maps out the difference in phase between the two waves, with maxima occurring when the phase difference is a multiple of 2π. If the two beams are of equal intensity, the maxima are four times as bright as the individual beams, and the minima have zero intensity.

The two waves must have the same polarization to give rise to interference fringes since it is not possible for waves of different polarizations to cancel one another out or add together. Instead, when waves of different polarization are added together, they give rise to a wave of a different polarization state.

Light Source Requirements

The discussion above assumes that the waves which interfere with one another are monochromatic, i.e. have a single frequency—this requires that they are infinite in time. This is not, however, either practical or necessary. Two identical waves of finite duration whose frequency is fixed over that period will give rise to an interference pattern while they overlap. Two identical waves which consist of a narrow spectrum of frequency waves of finite duration, will give a series of fringe patterns of slightly differing spacings, and provided the spread of spacings is significantly less than the average fringe spacing, a fringe pattern will again be observed during the time when the two waves overlap.

Conventional light sources emit waves of differing frequencies and at different times from different points in the source. If the light is split into two waves and then re-combined, each individual light wave may generate an interference pattern with its other half, but the individual fringe patterns generated will have different phases and spacings, and normally no overall fringe pattern will be observable. However, single-element light sources, such as sodium- or mercury-vapor lamps have emission lines with quite narrow frequency spectra. When these are spatially and colour filtered, and then split into two waves, they can be superimposed to generate interference fringes. All interferometry prior to the invention of the laser was done using such sources and had a wide range of successful applications.

A laser beam generally approximates much more closely to a monochromatic source, and it is much more straightforward to generate interference fringes using a laser. The ease with which interference fringes can be observed with a laser beam can sometimes cause problems in that stray reflections may give spurious interference fringes which can result in errors.

Normally, a single laser beam is used in interferometry, though interference has been observed using two independent lasers whose frequencies were sufficiently matched to satisfy the phase requirements.

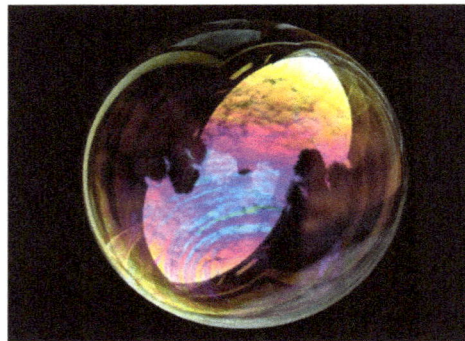

White light interference in a soap bubble

It is also possible to observe interference fringes using white light. A white light fringe pattern can be considered to be made up of a 'spectrum' of fringe patterns each of slightly different spacing. If all the fringe patterns are in phase in the centre, then the fringes will increase in size as the wavelength decreases and the summed intensity will show three to four fringes of varying colour. Young describes this very elegantly in his discussion of two slit interference. Some fine examples of white light fringes can be seen here. Since white light fringes are obtained only when the two waves have travelled equal distances from the light source, they can be very useful in interferometry, as they allow the zero path difference fringe to be identified.

Optical Arrangements

To generate interference fringes, light from the source has to be divided into two waves which have then to be re-combined. Traditionally, interferometers have been classified as either amplitude-division or wavefront-division systems.

In an amplitude-division system, a beam splitter is used to divide the light into two beams travelling in different directions, which are then superimposed to produce the interference pattern. The Michelson interferometer and the Mach-Zehnder interferometer are examples of amplitude-division systems.

In wavefront-division systems, the wave is divided in space—examples are Young's double slit interferometer and Lloyd's mirror.

Interference can also be seen in everyday phenomena such as iridescence and structural coloration. For example, the colours seen in a soap bubble arise from interference of light reflecting off the front and back surfaces of the thin soap film. Depending on the thickness of the film, different colours interfere constructively and destructively.

Applications

Optical Interferometry

Interferometry has played an important role in the advancement of physics, and also has a wide range of applications in physical and engineering measurement.

Thomas Young's double slit interferometer in 1803 demonstrated interference fringes when two small holes were illuminated by light from another small hole which was illuminated by sunlight. Young was able to estimate the wavelength of different colours in the spectrum from the spacing of the fringes. The experiment played a major role in the general acceptance of the wave theory of light. In quantum mechanics, this experiment is considered to demonstrate the inseparability of the wave and particle natures of light and other quantum particles (wave–particle duality). Richard Feynman was fond of saying that all of quantum mechanics can be gleaned from carefully thinking through the implications of this single experiment.

The results of the Michelson–Morley experiment are generally considered to be the first strong evidence against the theory of a luminiferous aether and in favor of special relativity.

Interferometry has been used in defining and calibrating length standards. When the metre was defined as the distance between two marks on a platinum-iridium bar, Michelson and Benoît used interferometry to measure the wavelength of the red cadmium line in the new standard, and also showed that it could be used as a length standard. Sixty years later, in 1960, the metre in the new SI system was defined to be equal to 1,650,763.73 wavelengths of the orange-red emission line in the electromagnetic spectrum of the krypton-86 atom in a vacuum. This definition was replaced in 1983 by defining the metre as the distance travelled by light in vacuum during a specific time interval. Interferometry is still fundamental in establishing the calibration chain in length measurement.

Interferometry is used in the calibration of slip gauges (called gauge blocks in the US) and in coordinate-measuring machines. It is also used in the testing of optical components.

Radio Interferometry

In 1946, a technique called astronomical interferometry was developed. Astronomical radio interferometers usually consist either of arrays of parabolic dishes or two-dimensional arrays of omni-directional antennas. All of the telescopes in the array are widely separated and are usually connected together using coaxial cable, waveguide, optical fiber, or other type of transmission line. Interferometry increases the total signal collected, but its primary purpose is to vastly increase the resolution through a process called Aperture synthesis. This technique works by superposing (interfering) the signal waves from the different telescopes on the principle that waves that coincide with the same phase will add to each other while two waves that have opposite phases will cancel each other out. This creates a combined telescope that is equivalent in resolution (though not in sensitivity) to a single antenna whose diameter is equal to the spacing of the antennas furthest apart in the array.

The Very Large Array, an interferometric array formed from many smaller telescopes, like many larger radio telescopes.

Acoustic Interferometry

An acoustic interferometer is an instrument for measuring the physical characteristics of sound wave in a gas or liquid. It may be used to measure velocity, wavelength, absorption, or impedance. A vibrating crystal creates the ultrasonic waves that are radiated into the medium. The waves strike a reflector placed parallel to the crystal. The waves are then reflected back to the source and measured.

Quantum Interference

If a system is in state , its wavefunction is described in Dirac or bra–ket notation as:

$$|\psi\rangle = \sum_i |i\rangle\langle i\,|\,\psi\rangle = \sum_i |i\rangle \psi_i$$

where the s specify the different quantum "alternatives" available (technically, they form an eigenvector basis) and the are the probability amplitude coefficients, which are complex numbers.

The probability of observing the system making a transition or quantum leap from state to a new state is the square of the modulus of the scalar or inner product of the two states:

$$\text{prob}(\psi \Rightarrow \varphi) = |\langle \psi\,|\,\varphi\rangle|^2 = \left|\sum_i \psi_i^* \varphi_i\right|^2$$

$$= \sum_{ij} \psi_i^* \psi_j \varphi_j^* \varphi_i = \sum_i |\psi_i|^2 |\varphi_i|^2 + \sum_{ij;\,i \neq j} \psi_i^* \psi_j \varphi_j^* \varphi_i$$

where (as defined above) and similarly are the coefficients of the final state of the system. * is the complex conjugate so that , etc.

Now let's consider the situation classically and imagine that the system transited from to via an intermediate state . Then we would classically expect the probability of the two-step transition to be the sum of all the possible intermediate steps. So we would have

$$\text{prob}(\psi \Rightarrow \varphi) = \sum_i \text{prob}(\psi \Rightarrow i \Rightarrow \varphi)$$

$$= \sum_i |\langle \psi \mid i \rangle|^2 |\langle i \mid \varphi \rangle|^2 = \sum_i |\psi_i|^2 |\varphi_i|^2,$$

The classical and quantum derivations for the transition probability differ by the presence, in the

quantum case, of the extra terms $\sum_{ij;i\neq i}\psi_i^*\psi_j\varphi_j^*\varphi_i$; these extra quantum terms represent interference between the different intermediate "alternatives". These are consequently known as the quantum interference terms, or cross terms. This is a purely quantum effect and is a consequence of the non-additivity of the probabilities of quantum alternatives.

The interference terms vanish, via the mechanism of quantum decoherence, if the intermediate state is measured or coupled with the environment.

References

- Morfey, Christopher L. (2001). Dictionary of Acoustics. San Diego: Academic Press. ISBN 978-0125069403.

- Winer, Ethan (2013). "1". The Audio Expert. New York and London: Focal Press. ISBN 978-0-240-82100-9.

- Greene, Brian (1999). The Elegant Universe: Superstrings, Hidden Dimensions, and the Quest for the Ultimate Theory. New York: W.W. Norton. pp. 97–109. ISBN 0-393-04688-5.

- Montgomery, Christopher. "24/192 Music Downloads …and why they make no sense". xiph.org. Retrieved 2016-03-17. The very quietest perceptible sound is about -8dbSPL

- "Sound Pressure is the force of sound on a surface area perpendicular to the direction of the sound". Retrieved 22 April 2015.

- Wolfe, J. "What is acoustic impedance and why is it important?". University of New South Wales, Dept. of Physics, Music Acoustics. Retrieved 1 January 2014.

- "EPA Identifies Noise Levels Affecting Health and Welfare" (Press release). Environmental Protection Agency. April 2, 1974. Retrieved October 17, 2014.

- Swanepoel, De Wet; Hall III, James W; Koekemoer, Dirk (February 2010). "Vuvuzela – good for your team, bad for your ears" (PDF). South African Medical Journal. 100 (4): 99–100. PMID 20459912.

Permissions

Index